Principles of Applied Mathematics

Principles of Applied Mathematics

Ada Hastings

MURPHY & MOORE
www.murphy-moorepublishing.com

www.murphy-moorepublishing.com

MURPHY & MOORE

Cataloging-in-publication Data

Principles of applied mathematics / Ada Hastings.
p. cm.
Includes bibliographical references and index.
ISBN 978-1-63987-720-1
1. Mathematics. I. Hastings, Ada.
QA36 .A67 2023
510--dc23

Murphy & Moore Publishing
1 Rockefeller Plaza,
New York City,
NY 10020, USA

ISBN 978-1-63987-720-1

Contents

Preface

Applied mathematics is an interdisciplinary field that emphasizes on the creation and imaginative use of mathematical concepts to pose and solve problems spanning physical and biomedical sciences to engineering and social sciences. The study of applied mathematics seeks to quantitatively describe, predict, design, and control phenomena in a range of fields such as physics, engineering, medicine, biology, finance, business, computer science, and industry. It examines the relations between models and observations and at the same time scrutinizes the mathematical foundations and limitations of these models and techniques. The field of applied mathematics is associated with scientific computing, computer science, operations research and management science, statistics, actuarial science, and mathematical economics. This book provides comprehensive insights into the principles of applied mathematics. It presents this complex subject in the most comprehensible and easy to understand language. This book aims to serve as a resource guide for graduate students studying applied mathematics.

This book has been the outcome of endless efforts put in by authors and researchers on various issues and topics within the field. The book is a comprehensive collection of significant researches that are addressed in a variety of chapters. It will surely enhance the knowledge of the field among readers across the globe.

It gives us an immense pleasure to thank our researchers and authors for their efforts to submit their piece of writing before the deadlines. Finally in the end, I would like to thank my family and colleagues who have been a great source of inspiration and support.

Ada Hastings

1

Probability and Statistics

1.1 Probability

$$\text{probability} = \frac{\text{possible choices}}{\text{total number of options}}$$

It is a branch of mathematics that deals with calculating the likelihood of a given event's occurrence, which is expressed as a number between 1 and 0. An event with a probability of 1 can be considered a certainty: for example, probability of a coin toss resulting in either "heads" or "tails" is 1, because there are no other options, assuming the coin lands flat. An event with a probability of .5 can be considered to have equal odds of occurring or not occurring: for example, probability of a coin toss resulting in "heads" is .5, because the toss is equally as likely to result in "tails". An event with a probability of 0 can be considered impossibility: for example, probability that the coin will land (flat) without either side facing up is 0, because either "heads" or "tails" must be facing up. A little paradoxical, probability theory applies precise calculations to quantify uncertain measures of random events.

In its simplest form, the probability can be expressed mathematically as: the number of occurrences of a targeted event divided by number of occurrences plus the number of failures of occurrences (this adds up to the total of possible outcomes),

$$p(a) = p(a)/[p(a) + p(b)]$$

Calculating probabilities in a situation like a coin toss is straightforward, because the outcomes are mutually exclusive: either one event or the other must occur. Each coin toss is an independent event the outcome of one trial has no effect on subsequent ones. No matter how many consecutive times one side lands facing up, probability that it will do so at the next toss is always .5 (50-50). The mistaken idea that a number of consecutive results (six "heads" for example) makes it more likely that the next toss will result in a "tails" is known as the gambler's fallacy, one that has led to the downfall of many a bettor.

Probability theory had its start in 17th century, when two French mathematicians, Blaise Pascal and Pierre de Fermat carried on a correspondence discussing mathematical problems dealing with the games of chance. Contemporary applications of probability theory run the gamut of human inquiry and include aspects of computer programming, astrophysics, music, weather prediction and medicine.

1.1.1 Random Variables

Sample space: Consider an experiment of throwing a coin twice. The outcomes S = {HH, HT, TH, TT} constitute the sample space.

Random variable: In this sample space each of these outcomes can be associated with a number by specifying a rule of association. Such a rule of association is called a random variables.

It is usually written X, is a variable whose possible values are numerical outcomes of the random phenomenon. There are two types of random variables, discrete and continuous.

Example: Number of heads

We denote random variable by the letter (X, Y, etc.) and any particular value of the random variable by x or y,

$$S = \{HH,\ HT,\ TH,\ TT\}$$

$$X(S) = \{2,\ 1,\ 1,\ 0\}$$

Thus a random X can be the considered as a function that maps all elements in the sample space S into points on the real line. The notation X(S) = x means that x is the value associated with outcome S by the random variable X.

Discrete Random Variables: A discrete random variable is one which may take on only a countable number of distinct values such as 0,1,2,3,4,........ Discrete random variables are usually (but not necessarily) counts. If a random variable can take only a finite number of distinct values, then it must be discrete. Examples of discrete random variables include the number of children in a family, the Friday night attendance at a cinema, the number of patients in a doctor's surgery, the number of defective light bulbs in a box of ten.

The probability distribution of a discrete random variable is a list of probabilities associated with each of its possible values. It is also sometimes called the probability function or the probability mass function.

Suppose a random variable X may take k different values, with the probability that $X = x_i$ defined to be $P(X = x_i) = p_i$. The probabilities pi must satisfy the following,

1: $0 < p_i < 1$ for each i.

2: $p_1 + p_2 + \ldots + p_k = 1$.

Distribution Function of the Random Variable X or Cumulative Distribution of the Random Variable X

The distribution function of a random variable X defined in (-∞, ∞) is given by,

$$F(x) = P(X \leq x) = P\{s : X(s) \leq x\}$$

Note:

Let the random variable X takes values $x_1, x_2, \ldots, x_n$ with probabilities $P_1, P_2, \ldots, P_n$

And let $x_1 < x_2 < \ldots < x_n$

Then we have,

$$F(x) = P(X < x_1) = 0, -\infty < x < x,$$
$$F(x) = P(X < x_1) = 0, P(X < x_1) + P(X = x_1) = 0 + p_1 = p_1$$
$$F(x) = P(X < x_2) = 0, P(X < x_1) + P(X = x_1) + p(X = x_2) = p_1 + p_2$$

$$F(x) = P(X < x_n) = 0, P(X < x_1) + P(X = x_1) + \ldots + p(X = x_n)$$
$$= p_1 + p_2 + \ldots\ldots + p_n = 1$$

Properties of Distribution Functions

Property: 1 $P(a < X \leq b) = F(b) - F(a)$, where $F(x) = P(X \leq x)$

Property: 2 $P(a \leq X \leq b) = P(X = a) + F(b) - F(a)$

Property: 3 $P(a < X < b) = P(a < X \leq b) - P(X = b)$

$$= F(b) - F(a) - P(X = b)$$

Example: Suppose a variable X can take the values 1, 2, 3 or 4. Probabilities associated with each outcome are described by following table,

Outcome1234	Probability0.10.30.40.2

The probability that X is equal to 2 or 3 is the sum of the two probabilities: $P(X = 2 \text{ or } X = 3) = P(X = 2) + P(X = 3) = 0.3 + 0.4 = 0.7$. Similarly, the probability that X is greater than 1 is equal to $1 - P(X = 1) = 1 - 0.1 = 0.9$, by the complement rule.

This distribution may also be described by probability histogram shown to the right.

All random variables (discrete and continuous) have a cumulative distribution function. It is a function giving probability that the random variable X is less than or equal to x, for every value x. For a discrete random variable, cumulative distribution function is found by summing up probabilities.

Example:

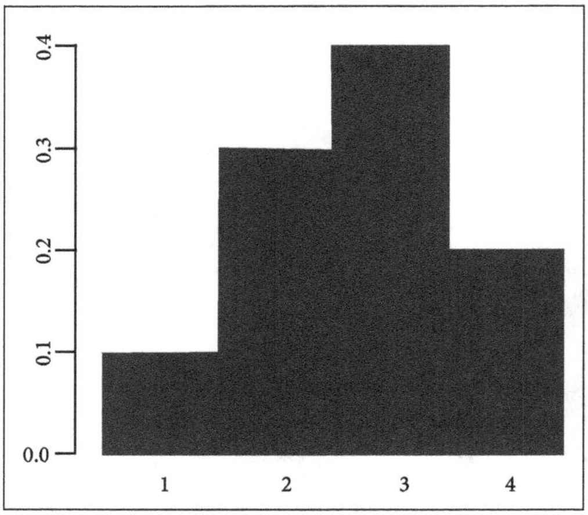

The cumulative distribution function for the above probability distribution is calculated as follows: Probability that X is less than or equal to 1 is 0.1, Probability that X is less than or equal to 2 is 0.1+0.3 = 0.4, Probability that X is less than or equal to 3 is 0.1 + 0.3 + 0.4 = 0.8 and Probability that X is less than or equal to 4 is 0.1 + 0.3 + 0.4 + 0.2 = 1.

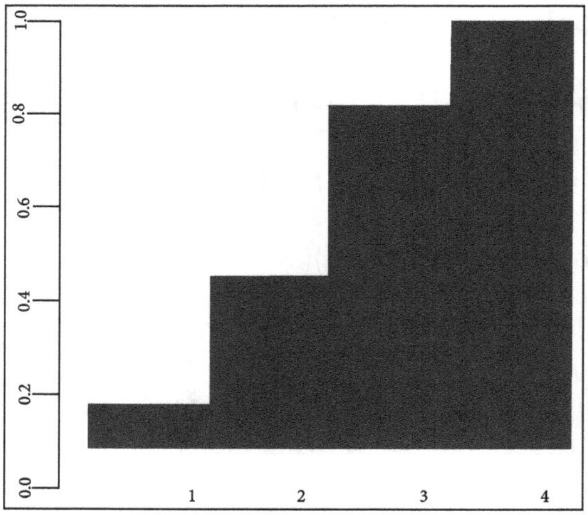

The probability histogram for the cumulative distribution of this random variable is shown to the right.

Continuous Random Variables

A continuous random variable is one which takes an infinite number of possible values.

Continuous random variables are usually measurements. Examples include height, weight, the amount of sugar in an orange, the time required to run a mile.

Suppose a random variable X may take all values over an interval of real numbers then the probability that X is in the set of outcomes A, P(A), is defined to be the area above A and under a curve. The curve, which represents a function p(x), must satisfy the following:

- The curve has no negative values $(p(x) > 0 \text{ for all } x)$.
- The total area under the curve is equal to 1.

If f(x) is a p.d.f. of a continuous random variable 'X', then the function

$$Fx(x) = F(x) = P(X \le x) = f(x)dx, x$$

$$Fx(x) = F(x) = P(X \le x) = \int_{-\infty}^{\infty} f(x)dx, -\infty < x < \infty$$

Properties of CDF of a random variable 'X':

- $0 \le F(x) \le 1, -\infty < x < \infty$
- $\underset{x \to -\infty}{\text{Lt}} F(x) = 0. \quad \underset{x \to -\infty}{\text{Lt}} F(x) = 1.$
- $P(a \le X \le b) = \int_a^b f(x)dx = F(b) - F(a)$
- $F'(x) = \dfrac{df(x)}{dx} = f(x) \ge 0$
- $P(X = x_i) = F(x_i) - F(x_i - 1)$

Example: In the experiment of throwing a coin twice the sample space S is S = {HH, HT, TH, TT}. Let X be a random variable chosen such that X(S) = x (the number of heads).

Note: Any random variable whose only possible values are 0 and 1 is called a Bernoulli random variable.

A curve meeting these requirements is known as a density curve.

The Uniform Distribution

Random number generator acting over an interval of numbers (a, b) has a continuous distribution. Since any interval of numbers of equal width has an equal probability of being observed, the curve describing the distribution is a rectangle, with constant height across the interval and 0 heights elsewhere. Since the area under curve must be equal to 1, length of the interval determines the height of the curve.

The following graphs plot the density curves for a random number generators over the intervals (4, 5) (top left), (2,6) (top right), (5,5.5) (lower left) and (3,5) (lower right). The distributions corresponding to these curves are known as uniform distributions.

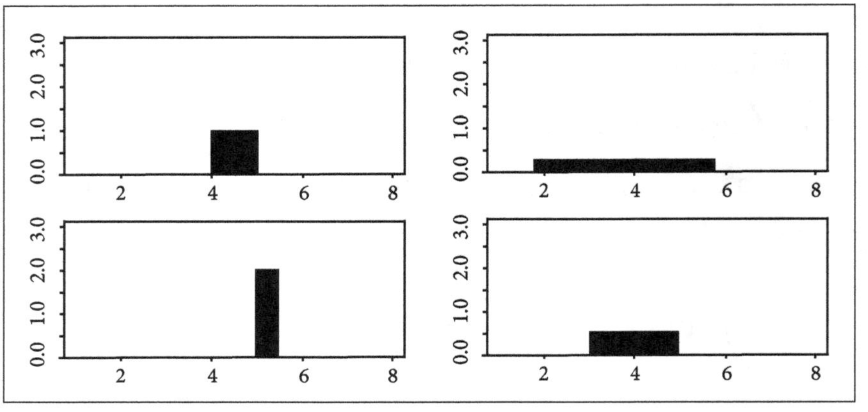

Consider the uniform random variable X defined on the interval (2, 6). Since the interval has width = 4, the curve has height = 0.25 over the interval and 0 elsewhere. Probability that X is less than or equal to 5 is the area between 2 and 5 or (5-2)*0.25 = 0.75. Probability that X is greater than 3 but less than 4 is the area between 3 and 4, (4-3)*0.25 = 0.25. To find that the probability that X is less than 3 or greater than 5, add the two probabilities(X < 3 and X > 5) = P(X < 3) + P(X > 5) = (3-2)*0.25 + (6-5)*0.25 = 0.25 + 0.25 = 0.5.

The uniform distribution is often used to simulate data. Suppose we would like to simulate data for 10 rolls of a regular 6-sided die. Using the MINITAB "RAND" command with "UNIF" sub-command generates 10 numbers in the interval (0, 6),

MTB > RAND 10 c2;

SUBC > unif 0 6.

Assign discrete random variable X to the values 1, 2, 3, 4, 5 or 6 as follows: if 0<X<1, X=1 if 1<X<2, X=2 if 2<X<3, X=3 if 3<X<4, X=4 if 4<X<5, X=5 if X>5, X=6. Use the generated MINITAB data to assign X to a value for each roll of the die:

Uniform data	X value
4.53786	5
5.77474	6
3.69518	4
1.03929	2
4.23835	5

0.37096	1
0.75272	1
5.56563	6
0.89045	1
3.18086	4

Another type of continuous density curve is the normal distribution. The area under the curve is not easy to calculate for a normal random variable X with mean μ and standard deviation σ. However, tables (and computer functions) are available for the standard random variable Z, which is computed from X by subtracting μ and dividing by σ. All of the rules of probability apply to the normal distribution.

Problems

1. A random variable X has the following probability function.

Values of x	0	1	2	3	4	5	6	7	8
Probability P(X)	a	3a	5a	7a	9a	11a	13a	15a	17a

- Let us determine the value of 'a'.
- Let us find $p(X<3), p(X\geq 3), p(0<X<5)$.
- Let us find the distribution function of X.

Solution:

Given:

Table 1:

Values of X	0	1	2	3	4	5	6	7	8
Probability P(X)	a	3a	5a	7a	9a	11a	13a	15a	17a

i. We know that if p(x) is the probability of mass function then,

$$\sum_{i=0}^{8} p(X_i) = 1$$

p(0) + p(1) + p(2) + p(3) + p(4) + p(5) + p(6) + p(7) + p(8) = 1

a + 3a + 5a + 7a + 9a + 11a + 13a + 15a + 17a = 1

81 a = 1

a = 1/81

Put a = 1/81 in table 1, get table 2.

Table 2:

X = x	0	1	2	3	4	5	6	7	8
P(x)	1/81	3/81	5/81	7/81	9/81	11/81	13/81	15/81	17/81

ii. $P(X < 3) = p(0) + p(1) + p(2)$

$= 1/81 + 3/81 + 5/81 = 9/81$

iii. $P(X \geq 3) = 1 - p(X < 3)$

$= 1 - 9/81 = 72/81$

iv. $P(0 < x < 5) = p(1) + p(2) + p(3) + p(4)$ here 0 and 5 are not include

$= 3/81 + 5/81 + 7/81 + 9/81$

v. To find the distribution function of X using table 2, we get,

X = x	$F(X) = P(x \leq x)$
0	$F(0) = p(0) = 1/81$
1	$F(1) = P(X \leq 1) = p(0) + p(1)$ $= 1/81 + 3/81 = 4/81$
2	$F(2) = P(X \leq 2) = p(0) + p(1) + p(2)$ $= 4/81 + 5/81 + 9/81$
3	$F(3) = P(X \leq 3) = p(0) + p(1) + p(2) + p(3)$ $= 9/81 + 7/81 + 16/81$
4	$F(4) = P(X \leq 4) = p(0) + p(1) + p(2) + ...p(4)$ $= 16/81 + 9/81 + 25/81$
5	$F(5) = P(X \leq 5) = p(0) + p(1) + p(2) + ...p(4) + p(5)$ $= 2/81 + 11/81 + 36/81$
6	$F(6) = P(X \leq 6) = p(0) + p(1) + p(2) + ...p(6)$ $= 36/81 + 13/81 + 49/81$
7	$F(7) = P(X \leq 7) = p(0) + p(1) + p(2) + ...p(6) + p(7)$ $= 49/81 + 15/81 + 64/81$
8	$F(8) = P(X \leq 8) = p(0) + p(1) + p(2) + ...p(7) + p(8)$ $= 64/81 + 17/81 + 81/81 = 1$

2. Let us find if Var(x) = 4, find Var(3x + 8),where X is a random variables.

Solution:

$$\text{Var}(x) = 4$$

$$\text{Var}(ax+b) = a^2 \text{ Varx}$$

$$\text{Var}(3x+8) = 3^2 \text{ Varx} = 36.$$

3. X and Y are independent random variables with variance 2 and 3. Let us determine the variance of 3X + 4Y.

Solution:

$$V(3X+4Y) = p\text{Var}(X) + 16\text{Var}(Y) + 24\text{Cov}(XY)$$

$$= 9*2+16*3+0(\therefore \text{X and Y are independent Cov}(XY)=0)$$

$$= 18+48 = 66.$$

4. Consider the cumulative distribution function of a R.V.X given by,

$$F(x) = \begin{cases} 1-\dfrac{4}{x^2}, & x>2 \\ 0, & x \le 2 \end{cases}$$

$$\text{Let us find } (1)p(x<3)(2)P(4<X<5)(3)P(X\ge 3).$$

Solution:

$$(1)\ P(x<3) = F(3) = 1-\frac{4}{3^2} = 1-\frac{4}{9} = \frac{5}{9}$$

$$(2)\ P(4<X<5) = F(\%) - F(4)$$

$$= \left(1-\frac{4}{25}\right) - \left(1-\frac{4}{16}\right)$$

$$= \frac{4}{16} - \frac{4}{25}$$

$$= \frac{9}{100} = 0.09$$

(3) $P(x \geq 3) = 1 - P(x < 3)$

$$= 1 - \frac{5}{9} = \frac{4}{9}$$

$$P(x \geq 3) = \frac{4}{9}.$$

5. Let X be a R.V. with E(X) = 1 and E(X(X - 1) = 4. Let us determine the $\text{Var}\left(\frac{x}{2}\right)$ and Var (2 - 3X).

Solution:

Given:

$$E(X) = 1; E[X(x-1)] = 4$$

i.e., $E[X^2] - E[x] = 4$

$$\Rightarrow E(X^2) - 1 = 4$$

$$\Rightarrow E(X^2) = 5$$

$$\text{Var}(x) = E(x^2) - (E(x))^2$$

$$= 5 - 1 = 4$$

$$\text{Var}(x) = 4.$$

Now,

$$\text{Var}\left(\frac{x}{2}\right) = \frac{1}{4}\text{Var}(x) \text{ W.K.T}$$

$$\text{Var}(aX + b) = s^2\text{Var}(X)$$

$$\text{Var}\left(\frac{x}{2}\right) = \frac{1}{4}(4) = 1$$

$$\text{Var}(2 - 3X) = 9\,\text{Var}(X)$$

$$= 9(4)$$

$$\text{Var}(2 - 3X) = 36$$

6. If X is a continuous R.V. with p.d.f. $f(x)=\begin{cases} x, & 0\le x<1 \\ \frac{3}{2}(x-1)^2, & 1\le x<2 \\ 0, & \text{otherwise.}\end{cases}$ Let us determine the cumulative distribution function F(x) of X and use it to find $P\left(\frac{3}{2}<X<\frac{5}{2}\right)$.

Solution:

Given:

$$f(x)=\begin{cases} x, & 0\le x<1 \\ \frac{3}{2}(x-1)^2, & 1\le x<2 \\ 0, & \text{otherwise}\end{cases}$$

We know that,

$$F(x)=\int_{-\infty}^{x} f(x)\,dx$$

When x lies in $0 \le x < 1$ we get,

$$F(x)=\int_{-\infty}^{x} f(x)dx=\int_{0}^{x} x\,dx=\left(\frac{x^2}{2}\right)_0^x \frac{x^2}{2}$$

When x lies in $1 \le x < 2$,

$$F(x)=\int_{-\infty}^{x} f(x)dx=\int_{0}^{0} f(x)dx+\int_{1}^{1} f(x)dx+\int_{2}^{x} f(x)dx$$

$$=\int_{0}^{1} x\,dx+\int_{1}^{x}\frac{3}{2}(x-1)^2dx$$

$$=\left(\frac{x^2}{2}\right)_0^1+\frac{3}{2}\left(\frac{(x-1)^3}{3}\right)_1^x$$

$$F(x)=\frac{1}{2}+\frac{1}{2}\left[(x-1)^3-(0)\right]$$

$$F(x)=\frac{1}{2}+\frac{(x-1)^3}{2}$$

When x lies in $x \ge 2$,

$$F(x)=\int_{-\infty}^{0} f(x)dx=\int_{-\infty}^{1} f(x)dx+\int_{0}^{2} f(x)dx+\int_{1}^{\infty} f(x)dx$$

$$=0+\int_0^1 x\,dx+\int_1^2 \frac{3}{2}(x-1)^2dx+0$$

$$=\left(\frac{x^2}{2}\right)_0^1+\frac{3}{2}\left[\frac{(x-1)^3}{3}\right]_1^2+0$$

$$=\frac{1}{2}+\frac{1}{2}$$

$$F(X)=1$$

∴ Cumulative distribution function F(X) of X

$$F(X)=\begin{cases} 0 & ; \quad x<0 \\ \dfrac{x^2}{2} & ; \quad 0\le x\le 1 \\ \dfrac{1}{2}+\dfrac{(x-1)^3}{2} & ; \quad 1\le x<2 \\ 1 & ; \quad x\ge 2 \end{cases}$$

$$P\left(3/2<X<\frac{5}{2}\right)=F(5/2)-F(3/2)$$

$$=1-\left(\frac{1}{2}+\frac{(0.5)^3}{2}\right)$$

$$=1-1/2-\frac{1}{16}$$

$$P\left(\frac{3}{2}<X<\frac{5}{2}\right)=\frac{7}{16}$$

7. Let X be a random variable with p.d.f. $f(x)=\frac{1}{\sqrt{2\pi}}e^{-x^2/2},-\alpha_1<x<\alpha_1$ Let us determine the p.d.f. of the R.V. $Y = X^2$.

Solution:

Given:

$$f(x)=\frac{1}{\sqrt{2\pi}}e^{-x^2/2},-\alpha_1<x<\alpha_1$$

$$F_Y(y)=P(Y\le y)=P(X^2\le y)$$

$$=P(-\sqrt{y}\le X\le\sqrt{y}) \text{ if } y\ge 0$$

$$F^{Y}(y) = Fx(\sqrt{y}) - F_x(-\sqrt{y}) \ldots (1)$$

And,

$$F_Y(y) = 0 \text{ if } y > 0$$

[Since $X^2 = Y$ has no roots, when $y > 0$].

Differentiating (1) with respect to y.

$$f_Y(y) = \frac{1}{2\sqrt{y}}\left\{f_X(\sqrt{y}) + f_X(-\sqrt{y})\right\}; \text{ if } y \geq 0$$

$$= 0 \text{ if } y < 0 \qquad \ldots (2)$$

$$f_X(x) = \frac{1}{\sqrt{2\pi}} e^{-x^2/2}; -\infty < x < \infty$$

$$\therefore f_X(\sqrt{y}) = \frac{1}{\sqrt{2\pi}} e^{-y/2} \text{ and } f_X(-\sqrt{y}) = \frac{1}{\sqrt{2\pi}} e^{-y/2}$$

Substituting in (2),

$$f_Y(y) = \frac{1}{2\sqrt{y}} - 2\frac{1}{\sqrt{2\pi}} e^{-y/2}$$

$$f_Y(y) = \frac{1}{\sqrt{2\pi y}} e^{-y/2}; y > 0.$$

8. If the density function of a continuous random variable X is given by,

$$F(x) = \begin{cases} ax & ; \quad 0 \leq x \leq 1 \\ a & ; \quad 1 \leq x \leq 2 \\ 3a - ax & ; \quad 2 \leq x \leq 3 \\ 0 & ; \quad \text{otherwise} \end{cases}$$

i. Let us find 'a'.

ii. Let us also determine the p.d.f of X.

Solution:

i. Since f(x) is a p.d.f, $\int_{R_X} f(x)dx = 1$

$$\text{i.e., } \int_0^1 f(x)dx = 1$$

$$\text{i.e., } \int_0^1 ax\,dx + \int_1^2 a\,dx + \int_2^3 (3a - ax)\,dx = 1$$

$$\text{i.e., } 2a = 1$$

$$\therefore\ a = \frac{1}{2}$$

ii. $F(x) = P(X \le x) = 0$, when $x < 0$

$$F(x) = \int_0^x \frac{x}{2}dx = \frac{x^4}{4}, \text{ when } 0 \le x \le 1$$

$$= \int_0^1 \frac{x}{2} + dx + \int_1^x \frac{1}{2}dx$$

$$= \frac{x}{2} - \frac{1}{4} \text{ when } 1 \le x \le 2$$

$$= \int_0^1 \frac{x}{2}dx + \int_1^2 \frac{1}{2}dx + \int_2^x \left(\frac{3}{2} - \frac{x}{2}\right)dx$$

$$= \frac{3}{2}x - \frac{x^2}{4} - \frac{5}{4}, \text{ when } 2 \le x \le 3$$

$$= 1, \text{ when } x > 3.$$

1.1.2 Probability Distributions

It is a table or an equation that links each outcome of a statistical experiment with its probability of occurrence.

Probability Distribution Prerequisites

To understand probability distributions, it is important to understand the variables, random variables and some notation:

- A variable is a symbol (A, B, x, y, etc.) that can take on any of a specified set of values.

- When the value of a variable is the outcome of a statistical experiment, that variable is a random variable.
- Normally, statisticians use a capital letter to represent a random variable and a lower-case letter, to represent one of its values. For example:
    - X represents the random variable X.
    - P(X) represents the probability of X.
    - $P(X = x)$ refers to the probability that the random variable X is equal to a particular value, denoted by x. As an example, $P(X = 1)$ refers to the probability that the random variable X is equal to 1.

Probability Distributions

An example will make clear relationship between random variables and probability distributions. Suppose we flip a coin two times. This simple statistical experiment can have four possible outcomes: HH, HT, TH and TT. Now, let the variable X represent the number of Heads that result from this experiment. The variable X can take on the values 0, 1 or 2. In this example, X is a random variable, because its value is determined by the outcome of a statistical experiment.

Consider the coin flip experiment described above. The table below, which associates each outcome with its probability, is an example of a probability distribution:

Number of heads	Probability
0	0.25
1	0.50
2	0.25

The above table represents probability distribution of the random variable X.

Cumulative Probability Distributions

It refers to the probability that the value of a random variable falls within a specified range.

If we flip a coin two times, what is the probability that the coin flips would result in one or fewer heads? The answer would be a cumulative probability. It would be the probability that the coin flip experiment results in zero heads plus the probability that the experiment results in one head.

$$P(X < 1) = P(X = 0) + P(X = 1) = 0.25 + 0.50 = 0.75$$

Like a probability distribution, a cumulative probability distribution can be represented by a table or an equation. In the table below, the cumulative probability refers to the probability than the random variable X is less than or equal to x.

Number of heads:x	Probability:P(X=x)	Cumulative Probability:P(X<x)
0	0.25	0.25
2	0.50	0.75
2	0.25	1.00

Uniform Probability Distribution

The simplest probability distribution occurs when all of the values of a random variable occur with equal probability. Uniform probability distribution is called the uniform distribution.

Suppose the random variable X can assume k different values. Suppose also that the $P(X = xk)$ is constant. Then,

$$P(X = xk) = 1/k$$

Example 1: Let us determine the probability of the die landing on a 5 when it is tossed.

Solution:

When a die is tossed, there are 6 possible outcomes represented by: $S = \{1, 2, 3, 4, 5, 6\}$. Each possible outcome is a random variable (X) and each outcome is equally likely to occur. Thus, we have a uniform distribution. Therefore, the $P(X = 5) = 1/6$.

Example 2: Consider the same case as in Example 1, here let us determine the probability of the die landing on a number that is smaller than 5.

Solution:

When a die is tossed, there are 6 possible outcomes represented by: $S = \{1, 2, 3, 4, 5, 6\}$. Each possible outcome is equally likely to occur. Thus, we have a uniform distribution.

This problem involves a cumulative probability. The probability that the die will land on a number smaller than 5 is equal to:

$$P(X < 5) = P(X = 1) + P(X = 2) + P(X = 3) + P(X = 4)$$

$$= 1/6 + 1/6 + 1/6 + 1/6 = 2/3$$

A probability distribution is a table or an equation that links each outcome of a statistical experiment with its probability of occurrence.

1.2 Mean and Variance of Distribution

Mean

The mean of the sampling distribution is the mean of the population from which scores were sampled. Therefore, if a population has a mean μ, then the mean of the sampling distribution of the mean is also μ. Symbol **μM** is used to refer to the mean of the sampling distribution of the mean. Therefore, the formula for mean of the sampling distribution of mean can be written as,

$$\mu_M = \mu$$

Variance

The variance of the sampling distribution of mean is computed as follows,

$$\sigma_M^2 = \frac{\sigma^2}{N}$$

That is, the variance of the sampling distribution of the mean is the population variance divided by N, the sample size (the number of scores used to compute a mean). Thus, the larger the sample size, the smaller the variance of the sampling distribution of the mean.

This expression can be derived very easily from variance sum law. Let's begin by computing the variance of the sampling distribution of the sum of three numbers sampled from a population with variance σ^2. The variance of the sum would be $\sigma^2 + \sigma^2 + \sigma^2$. For N numbers, the variance would be Nσ2. Since the mean is 1/N times the sum, the variance of the sampling distribution of the mean would be $1 / N^2$ times the variance of the sum, which equals σ^2 / N.

The standard error of the mean is standard deviation of the sampling distribution of the mean. It is therefore the square root of the variance of the sampling distribution of mean and can be written as,

$$\sigma_M = \frac{\sigma}{\sqrt{N}}$$

The standard error is represented by a σ because it is a standard deviation. The subscript (M) indicates that the standard error in question is the standard error of the mean.

Central Limit Theorem

It states that:

Given a population with a finite mean μ and a finite non-zero variance σ^2, the sampling distribution of the mean approaches a normal distribution with a mean of μ and a variance of σ^2/N as N, the sample size, increases.

The expressions for the mean and variance of the sampling distribution of the mean are not new or remarkable. What is remarkable is that regardless of the shape of the parent population, the sampling distribution of the mean approaches a normal distribution as N increases. As a reminder, Figure shows the results of simulation for N = 2 and N = 10. The parent population was a uniform distribution. We can see that the distribution for N = 2 is far from a normal distribution. Nonetheless, it does show that the scores are denser in the middle than in the tails. For N = 10 the distribution is quite close to a normal distribution. Notice that the means of the two distributions are the same, but that the spread of the distribution for N = 10 is smaller.

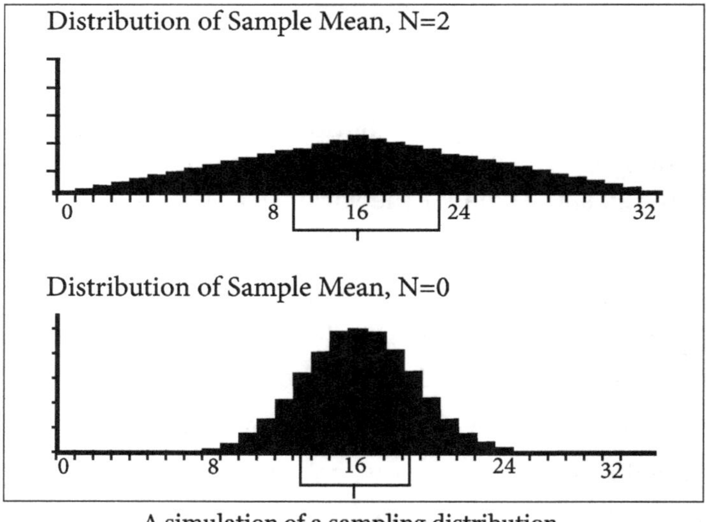

A simulation of a sampling distribution.

The parent population is uniform. The line under “16” indicates that 16 is the mean. The line extends from the mean plus and minus one standard deviation.

Figure shows how closely the sampling distribution of the mean approximates a normal distribution even when the parent population is very non-normal. If we look closely we can see that the sampling distributions do have a slight positive skew. The larger the sample size, the closer the sampling distribution of the mean would be to a normal distribution.

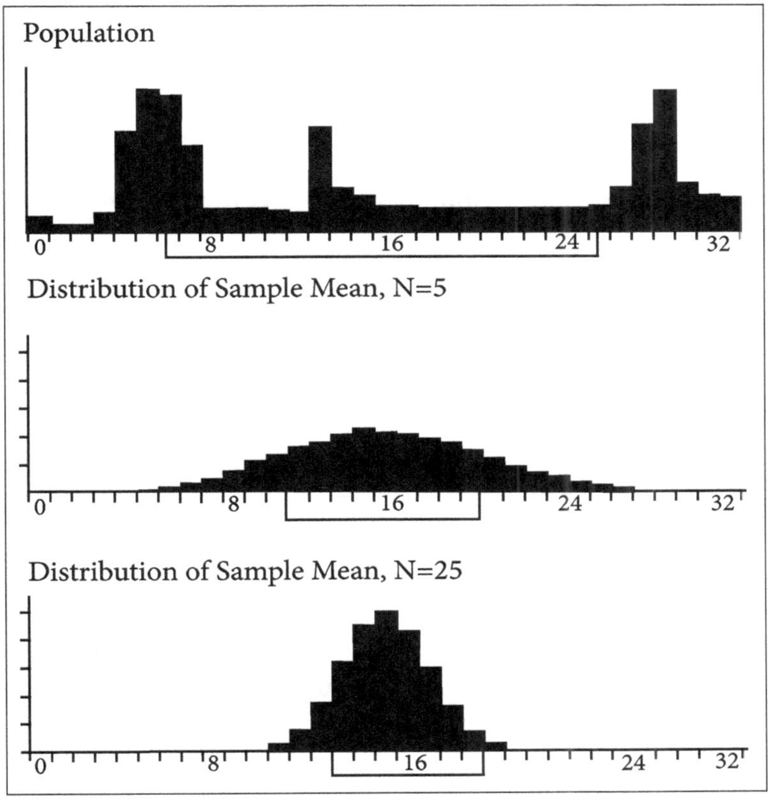

A simulation of a sampling distribution.

The parent population is very non-normal.

1.2.1 Binomial

A random variable X is said to follow binomial distribution if its probability law is given by,

$$P(x) = p(X = x \text{ successes}) = nC_x p_x q^{n-x}$$

Where,

$$x = 0, 1, 2, \ldots\ldots, n, \; p + q = 1$$

Note: Assumptions in Binomial distribution:

- There are only two possible outcomes for each trail (success or failure).
- The probability of a success is the same for each trail.
- There are 'n' trails, where 'n' is a constant.
- The 'n' trails are independent.

Theorem: Additive Property of Binomial Distribution

Proof:

The sum of two binomial variants is not a binomial variate.

Let X and Y be two independent binomial variates with parameter (n_1, p_1) and (n_2, p_2) respectively.

Then,

$$M_x(t) = (q_1 + p_1 e^t)^{n1},\ M_x(t) = (q_2 + p_2 e^t)^{n2}$$

$$\therefore M_{x+y}(t) = M_x(t).\, M_y(t) [\because \text{X and Y are independent R.V.'s}]$$

$$= (q_1 + p_1 e^t)^{n1} \cdot (q_2 + p_2 e^t)^{n2}$$

RHS cannot be expressed in the form $(q + pe^t)^n$. Hence by uniqueness theorem of MGF X + Y is not a binomial variate. Hence in general, the sum of two binomial variates is not a binomial variate.

Problems

1. Let us determine the Moment Generating Function (MGF) of a binomial distribution about origin.

Solution:

Given:

$$\text{WKT} \quad M_X(t) = \sum_{x=0}^{n} e^{tx} p(x)$$

Let 'X' be a random variable which follows the binomial distribution then MGF about origin is given by,

$$E[e^{tx}] = M_X(t) = \sum_{x=0}^{n} e^{tx} p(x)$$

$$= \sum_{x=0}^{n} e^{tx}\, nC_x p^x q^{n-x} \qquad [\therefore p(x) = nC_x\, p^x q^{n-x}]$$

$$= \sum_{x=0}^{n} (e^{tx}) p^x nC_x q^{n-x}$$

$$=\sum_{x=0}^{n}\left(pe^{t}\right)^{x} nC_{x} q^{n-x}$$

$$M_{x}(t)=\left(q+pe^{t}\right)^{n}$$

2. In a binomial distribution, the mean is 4 and variance is 3, let us determine P(X=0).

Solution:

Given:

Mean = 4

Variance = 3

Formula to be used:

$$P(X = x) = nc_{x}p^{x}q^{n-x}$$

$$np=4,\ npq=3$$

Hence,

$$q=3/4,\ p=1-q=1/4,$$

Since,

$$np=4,\ n=16.$$

$$P(X=x)=nc_{x}p^{x}q^{n-x}$$

$$=16c_{x}\left(\frac{1}{4}\right)^{x}\left(\frac{3}{4}\right)^{16-x} \quad x=0,1,2,...$$

$$P(X=0)=\left(\frac{3}{4}\right)^{16}=0.01002$$

3. If X is binomially distributed with n = 6 such that P(X = 2) = 9P(X = 4), let us determine E(x) and Var(x).

Solution:

Given:

$$n=6$$
$$P(X=2)=9P(X=4)$$

$6C2p2q4 = 9(6C4p4q2)$

$q = 3p,\ p = 1/4.$

$E(X) = 1.5;\ Var(X) = 9/8$

4. Let us determine the mean and variance of binomial distribution.

Solution:

$$M_x(t) = (q + pe^t)^n$$

$$\therefore\ Mx(t) = n(q + pe^t)^{n-1}.pe^t$$

Put,

t = 0, we get

$$M'_x(0) = n(q+p)^{n-1}.p$$

$$Mean = E(x) = np\ [\because (q+p) = 1]\ [Mean\ M'_x(0)]$$

$$M''_x(t) = np\left[(q+pe^t)^{n-1}.e^t + e^t(n-1)(q+pe^t)^{n-2}.pe^t\right]$$

Put,

t = 0, we get,

$$M''_x(t) = np\left[(q+p)^{n-1} + (n-1)(q+p)^{n-2}.p\right]$$

$$= np\left[1 + (n-1)p\right]$$

$$= np + n^2p^2 - np^2$$

$$= n^2p^2 - np(1-p)$$

$$M''_x(0) = n^2p^2 + npq\ [\because\ 1 - p = q]$$

$$M''_x(0) = E(X^2) = n^2p^2 + npq$$

$$Var(X) = E(X^2) - [E(X)]^2 = n^2/p^2 + npq - n^2/p^2 = npq$$

$$Var(X) = npq$$

$$S.D = \sqrt{npq}$$

1.2.2 Poisson

Poisson distribution is a discrete probability distribution, which is the limiting case of the binomial distribution under certain conditions.

When n is very indefinitely very large.

Probability of success is very small.

$$np = \lambda \text{ is finite, } \lambda \in R^{+}$$

Definition: A discrete random variable X is said to be follow a Poisson distribution if the probability mass function is given by,

Examples where the Poisson distribution is used (or) Applications of Poisson distribution.

This distribution is used to describe the behavior of the rare events like:

- The number of blind born per year in a large city.
- The number of printing mistakes per page in a large volume of a book.
- The number of air pockets in a glass sheet.
- The number of accidents occurred annually at a busy crossing of city.
- The number of defective articles produced by a quality machine.
- This is widely used in waiting lines or queuing problems in management studies.
- It has wide applications in industrial quality control.
- In determining the number of deaths in a given period by a rare disease.

The sum of the probabilities of the Poisson distribution is unity i.e. $\sum_{x=0}^{\infty} P(x^{-},\lambda) = 1.$

Proof:

For a Poisson distribution the probability mass function is given by,

$$p(X = x) = P(x^{-},\lambda) = \frac{e^{-}\lambda^{x}}{x!}, x = 0,1,2,3.........\infty$$

Now,

$$\sum_{x=0}^{\infty} P(x^{-},\lambda) = \sum_{x=0}^{\infty} \frac{e^{-\lambda}\lambda}{x!}$$

$$= e^{-\lambda}\sum_{x=0}^{\infty} \frac{\lambda^{x}}{x!}$$

$$= e^{-\lambda}\left[\frac{\lambda^{0}}{0!} + \frac{\lambda^{1}}{1!} + \frac{\lambda^{2}}{2!} + \frac{\lambda^{3}}{3!} \ldots\ldots\ldots + \infty\right]$$

$$= e^{-\lambda} e^{\lambda}$$

$$= 1$$

Mean and Variance of Poisson Distribution

For a Poisson distribution the probability mass function is given by,

$$p(X = x) = P(x^{-},\lambda^{x}) = \frac{e^{-\lambda}\lambda^{x}}{x!}, x = 0,1,2,3\ldots\ldots\ldots\infty$$

Now,

$$\text{Mean} = E(X) = \sum_{x=0}^{\infty} x P(x^{-},\lambda) = \sum_{x=0}^{\infty} x \frac{e^{-\lambda}\lambda^{x}}{x!}$$

$$= e^{-\lambda}\lambda\sum_{x=0}^{\infty} x \frac{\lambda\lambda^{x-1}}{x(x-1)!}$$

$$= e^{-\lambda}\lambda\sum_{x=1}^{\infty} \frac{\lambda^{x-1}}{(x-1)!}$$

$$= e^{-\lambda}\lambda\left[\frac{\lambda^{0}}{0!} + \frac{\lambda^{1}}{1!} + \frac{\lambda^{2}}{2!} + \frac{\lambda^{3}}{3!} \ldots\ldots\ldots + \infty\right]$$

$$= e^{-\lambda}e^{\lambda}$$

$$= \lambda$$

$\therefore$ Mean of the Poisson distribution is λ.

$$\text{Variance} = V(X) = E(X^{2}) = [E(x)]^{2}$$

$$= E(X^{2}) = [\lambda]^{2} \qquad \ldots(1)$$

[$\therefore$ Mean of the Poisson distribution is λ]

If from (1) ,

$$E(X^2) = \sum_{x=0}^{\infty} x^2 P(x^-, \lambda) = \sum_{x=0}^{\infty} x^2 \frac{e^{-\lambda} \lambda^x}{x!} \lambda$$

$$= \sum_{x=0}^{\infty} [x(x-1) + x] \frac{e^{-\lambda} \lambda^x}{x!}$$

$$= \sum_{x=0}^{\infty} x(x-1) \frac{e^{-\lambda} \lambda^x}{x!} + \sum_{x=0}^{\infty} x \frac{e^{-\lambda} \lambda^x}{x!}$$

$$= e^{-\lambda} \sum_{x=1}^{\infty} x(x-1) \frac{\lambda^2 \lambda^{x-2}}{x(x-1)(x-2)!} + \lambda$$

[$\therefore$ Mean of the Poisson distribution is λ]

$$= e^{-\lambda} \lambda^2 \sum_{x-2}^{\infty} \frac{\lambda^{x-2}}{(x-2)!} + \lambda$$

$$= e^{-\lambda} \lambda^2 \left[\frac{\lambda^0}{0!} + \frac{\lambda^1}{1!} + \frac{\lambda^2}{2!} + \frac{\lambda^3}{3!} \dots\dots\dots + \infty \right] + \lambda$$

$$= e^{-\lambda} \lambda^2 e^{\lambda} + \lambda$$

$$\Rightarrow E(X^2) = \lambda^2 + \lambda \qquad \dots(2)$$

Putting (2) in (1) we get

$$V(X) = E(X^2) = [\lambda^2]$$

$$= \lambda^2 + \lambda - \lambda^2$$

$$= \lambda$$

$\therefore$ Variance of the Poisson distribution is λ.

Note: In Poisson distribution the mean and variance are equal i.e. λ.

Moments of the Poisson Distribution

Non central moments (about zero):

$$\mu_1^1 = \text{Mean} = E(X) = \lambda$$

$$\mu_2^1 = E(X^2) = \lambda^2 + \lambda$$

$$\mu_3^1 = E(X^3) = \lambda^3 + 3\lambda^2 + \lambda$$

$$\mu_4^1 = E\left(X^4\right) = \lambda^4 + 6\lambda^3 + 7\lambda^2 + \lambda$$

Central moments (about mean):

$$\mu_1 = 0$$
$$\mu_2 = \text{variance} = \lambda$$
$$\mu_3 = \lambda$$
$$\mu_4 = 3\lambda^2 + \lambda$$

Skew Ness of Poisson Distribution

Measure of Skew ness of Poisson distribution is given by,

$$\sqrt{\beta_1} = \sqrt{\frac{\mu_3^2}{\mu_2^3}} = \sqrt{\frac{\lambda^2}{\lambda^3}} = \sqrt{\frac{1}{\lambda}}$$

Kurtosis of Poisson Distribution

Measure of Kurtosis of Poisson distribution is given by,

$$\beta_2 = \frac{\mu_4}{\mu_2^2} = \frac{3\lambda^2 + \lambda}{\lambda^2} = 3 + \frac{1}{\lambda}$$

Note:

- Since Skew ness = $\sqrt{\frac{1}{\lambda}} > 0$, Poisson distribution is always positively skewed.
- Since Kurtosis = $3 + \frac{1}{\lambda} > 3$, Poisson distribution is always Leptokurtic.

Recursion formula:

If X is a Poisson variate with p.m.f $P(X = x) = \frac{e^{-\lambda}\lambda^x}{x!}, x = 0,1,2,3..........\infty$

And,

$$P(X = x-1) = \frac{e^{-\lambda}\lambda^{x-1}}{x!},\ x = 0,1,2,3..........\infty$$

Now,

$$\frac{P(X = x)}{P(X = x-1)} = \frac{\frac{e^{-\lambda}\lambda^x}{x!}}{\frac{e^{-\lambda}\lambda^{x-1}}{(x-1)!}} = \frac{\lambda}{x}$$

$$\Rightarrow P(X=x)=\frac{\lambda}{x}P(X=x-1)$$

Problem

Prove that the Poisson distribution is the limiting case of binomial distribution stating the required conditions.

Solution:

The Poisson distribution can be limiting case of a binomial distribution under certain conditions.

- Number of trails i.e. n is indefinitely large i.e. $n \rightarrow \infty$.
- p, the probability of success in each trail is indefinitely small i.e $p \rightarrow 0$.
- $np = \lambda$ is finite.

If X is a binomial variate then the probability mass function is given by,

$$P(X=x)={}^{n}c_{x}p^{x}q^{n-x},\ x=0,1,2,\ldots\ldots,n$$

Under the above conditions,

$$\lim_{n\to\infty} B(x^{-},n,p)=\lim_{n\to\infty} {}^{n}c_{x}\,p^{x}q^{n-x}$$

$$=\lim_{n\to\infty}\frac{n!}{x!(n-x)!}\left(\frac{\lambda}{n}\right)^{x}\left(1-\frac{\lambda}{n}\right)^{n-x}\quad [\because np=\lambda]$$

$$=\lim_{n\to\infty}\frac{n(n-1)(n-2)\ldots\ldots\ldots(n-x+1)(n-x)!}{x!(n-x)!}\left(\frac{\lambda}{n}\right)^{x}\left(1-\frac{\lambda}{n}\right)^{n-x}$$

$$=\frac{\lambda^{x}}{x!}\lim_{n\to\infty}\frac{n^{x}\left[1\left(1-\frac{1}{n}\right)\left(1-\frac{2}{n}\right)\ldots\ldots\ldots\left(1-\frac{x-1}{n}\right)\right]}{n^{x}}\frac{\left(1-\frac{\lambda}{n}\right)^{n}}{\left(1-\frac{\lambda}{n}\right)^{x}}$$

$$=\frac{\lambda^{x}}{x!}\lim_{n\to\infty}1.\frac{\left(1-\frac{\lambda}{n}\right)^{n}}{\left(1-\frac{\lambda}{n}\right)^{x}}$$

$$=\frac{\lambda^x}{x!}\frac{\lim\limits_{n\to\infty}\left(1-\frac{\lambda}{n}\right)^n}{\lim\limits_{n\to\infty}\left(1-\frac{\lambda}{n}\right)^x}=\frac{\lambda^x}{x!}\frac{e^{-\lambda}}{1^x}=\frac{e^{-\lambda}\lambda^x}{x!}$$

Mode of the Poisson Distribution

Case 1: When λ is not an integer, the distribution is unimodal and the value of mode is the integral part of the λ.

Case 2: When λ is an integer, the distribution is bimodal and the value of modes are λ and λ - 1.

Show that in a Poisson distribution with unit mean, mean deviation is 2/e times the standard deviation.

Solution:

The Poisson distribution with unit mean i.e. $\lambda = 1$ is given by,

$$P(x;\,1)=\frac{e^{-1}1^x}{x!}=\frac{e^{-1}}{x!}$$

Now, we have to show that,

$$\text{Mean deviation about mean} = \frac{2}{e}\ \text{Standard deviation}$$

$$\text{i.e. } E|X-E(X)|=\frac{2}{e}E(X^2)-[E(X)]^2$$

$$\text{i.e. } E|X-1|=\frac{2}{e}\sqrt{1} \qquad [\therefore \text{ given mean}=1]$$

$$\text{i.e. } E|X-1|=\frac{2}{e}$$

Now,

$$E|X-1|=\sum_{x-0}^{\infty}|x-1|\frac{e^{-1}}{x!}$$

$$=\frac{1}{e}\left[1+\sum_{x-1}^{\infty}\frac{x-1}{x!}\right]$$

$$=\frac{1}{e}\left[1+\sum_{x-1}^{\infty}\frac{1}{(x-1)!}-\sum_{x-1}^{\infty}\frac{1}{x!}\right]$$

$$= \frac{1}{e}\left[1+e-(e-1)\right]$$

$$= \frac{2}{e}$$

Hence in Poisson distribution with unit mean, mean deviation are 2/e times the standard deviation.

Properties of Poisson Distribution

The following are the some of the properties of the Poisson distribution:

- Poisson distribution is a discrete probability distribution with single parameter λ.
- Both mean and variance of the Poisson distribution are equal to λ.
- The distribution is positively skewed and leptokurtic.
- It is asymptotic form of binomial distribution when p is small, n is large and np is finite.
- The normal distribution is a limiting form of a Poisson distribution as $\lambda \to 0$.
- The distributioon of rare events generally approximates to a Poisson distribution.
- If X_1 and X_2 are two independent Poisson variates with mean $\lambda 1$ and $\lambda 2$ respectively, then $X = X_1 + X_2$ is also a Poisson variate with mean $\lambda_1 + \lambda_2$.

1.2.3 Hyper Geometric Distributions

The probability distribution of a hyper geometric random variable is called a hyper geometric distribution. This lesson describes how hypergeometric random variables, hyper geometric experiments, hyper geometric probability and the hypergeometric distribution are all related.

Notation

The following notation is helpful, when we talk about hyper geometric distributions and hyper geometric probability:

- N: The number of items in the population.

- k: The number of items in the population that are classified as successes.
- n: The number of items in the sample.
- x: The number of items in the sample that are classified as successes.
- Cx: The number of combinations of k things, taken x at a time.
- h(x; N, n, k): Hyper geometric probability - the probability that an n-trial hyper geometric experiment results in exactly x successes, when the population consists of N items, k of which are classified as successes.

Hyper Geometric Experiments

A hyper geometric experiment is a statistical experiment that has the following properties:

- A sample of size n is randomly selected without replacement from a population of N items.
- In the population, k items can be classified as successes and N - k items can be classified as failures.

Consider the following statistical experiment. We have 10 marbles - 5 red and 5 green. We randomly select 2 marbles without replacement and count the number of red marbles we have selected. This would be a hyper geometric experiment.

Note that it would not be a binomial experiment. A binomial experiment requires that the probability of success be constant on every trial. With the above experiment, the probability of a success changes on every trial. In the beginning, the probability of selecting a red marble is 5/10. If we select a red marble on the first trial, the probability of selecting a red marble on the second trial is 4/9 and if we select a green marble on the first trial, the probability of selecting a red marble on the second trial is 5/9.

Note further that if we selected the marbles with replacement, the probability of success would not change. It would be 5/10 on every trial. Then, this would be a binomial experiment.

Hyper Geometric Distribution

A hyper geometric random variable is the number of successes that result from a hyper geometric experiment. The probability distribution of a hyper geometric random variable is called a hyper geometric distribution.

Given x, N, n and k, we can compute the hyper geometric probability based on the following formula:

Hyper Geometric Formula

Suppose a population consists of N items, k of which are successes and a random sample drawn from that population consists of items, x of which are successes. Then the hyper geometric probability is,

$$h(x; N, n, k) = [kCx] [N\text{-}kCn\text{-}x]/[NCn]$$

The hyper geometric distribution has the following properties:

- The mean of the distribution is equal to n * k/N.
- The variance is $n * k * (N - k) * (N - n)/[N^2 * (N - 1)]$.

Problem

Example 1: Suppose we randomly select 5 cards without replacement from an ordinary deck of playing cards. What is the probability of getting exactly 2 red cards (i.e., hearts or diamonds)?

Solution:

This is a hyper geometric experiment in which we know the following:

- N = 52; since there are 52 cards in a deck.
- k = 26; since there are 26 red cards in a deck.
- n = 5; since we randomly select 5 cards from the deck.
- x = 2; since 2 of the cards we select are red.

We plug these values into the hyper geometric formula as follows,

$h(x; N, n, k) = [{}_kC_x] [{}_{Nk}C_{nx}]/[{}_NC_n]$ $h(2; 52, 5, 26) = [{}_{26}C_2] [{}_{26}C_3]/[{}_{52}C_5]$ $h(2; 52, 5, 26) = [325] [2600]/[2{,}598{,}960] = 0.32513$

Thus, the probability of randomly selecting 2 red cards is 0.32513.

Cumulative Hyper Geometric Probability

A cumulative hyper geometric probability refers to the probability that the hyper geometric random variable is greater than or equal to some specified lower limit and less than or equal to some specified upper limit.

For example, suppose we randomly select five cards from an ordinary deck of playing cards. We might be interested in the cumulative hyper geometric probability of obtaining 2 or fewer hearts. This would be the probability of obtaining 0 hearts plus the probability of obtaining 1 heart plus the probability of obtaining 2 hearts.

Example 2: Suppose we select 5 cards from an ordinary deck of playing cards. What is the probability of obtaining 2 or fewer hearts?

Solution:

This is a hyper geometric experiment in which we know the following:

- N = 52; since there are 52 cards in a deck.
- k = 13; since there are 13 hearts in a deck.
- n = 5; since we randomly select 5 cards from the deck.
- x = 0 to 2; since our selection includes 0, 1 or 2 hearts.

We plug these values into the hyper geometric formula as follows:

$h(x < x; N, n, k) = h(x < 2; 52, 5, 13)$

$h(x < 2; 52, 5, 13) = h(x = 0; 52, 5, 13) + h(x = 1; 52, 5, 13) + h(x = 2; 52, 5, 13)$

$h(x < 2; 52, 5, 13) = [({}_{13}C_0) ({}_{39}C_5)/({}_{52}C_5)] + [({}_{13}C_1) ({}_{39}C_4)/({}_{52}C_5)] + [({}_{13}C_2) ({}_{39}C_3)/({}_{52}C_5)]$

$h(x<2;52,5,13)=[(1)(575,757)/(2,598,960)] + [(13)(82,251)/(2,598,960)] + [(78)(9139)/(2,598,90]$

$h(x < 2; 52, 5, 13) = [0.2215] + [0.4114] + [0.2743]$

$h(x < 2; 52, 5, 13) = 0.9072$

1.3 Normal and Exponential Distribution

Normal Distribution

The normal distribution is defined by the following probability density function, where μ is the population mean and σ^2 is the variance,

$$f(x) = \frac{1}{\sigma\sqrt{2\pi}} e^{-(x-\mu)^2/2\sigma^2}$$

If a random variable X follows the normal distribution, then we write,

$$X \sim N(\mu, \sigma^2)$$

In particular, the normal distribution with $\mu = 0$ and $\sigma = 1$ is called the standard normal distribution and is denoted as N(0,1). It can be graphed as follows.

The normal distribution is important because of the Central Limit Theorem, which states that the population of all possible samples of size n from a population with mean μ and variance $\sigma 2$ approaches a normal distribution with mean μ and $\sigma 2/n$ when n approaches infinity.

Definition: Data are said to be normally distributed if the rate at which the frequencies fall off is proportional to the distance of the score from the mean and to the frequencies themselves.

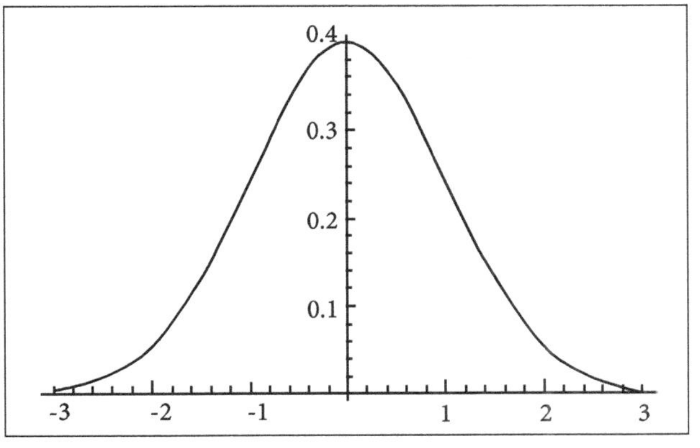

In practice, the value of the bell shaped curve is that we can find the proportion of the scores which lie over a certain interval. In a probability distribution, this is the area under the curve over the interval: a typical calculus problem. However, in order to use calculus to find these areas, we need a formula for the curve. We can find such a formula because our definition gives us the following differential equation,

$$\frac{df}{dx} = -k(x-\mu)f(x)$$

Where k is a positive constant. Note that to the right of the mean, the curve will be decreasing and to the left, it will be increasing. We can separate the variables,

$$\frac{df}{f} = -k(x-\mu)dx$$

And we can integrate both sides,

$$\int \frac{df}{f} = -k\int (x-\mu)dx$$

$$\text{In}\, f = \frac{-k(x-\mu)^2}{2} + \text{In}\, C$$

Take exponentials of both sides,

$$f(x)=Ce^{-\frac{k}{2}(x-\mu)^2}$$

We can use the fact that the normal distribution is a probability distribution and the total area under the curve is 1. If f(x) is a probability measure, then

$$c\int_{-\infty}^{\infty} e^{-\frac{k}{2}(x-\mu)^2}\,dx=1$$

It is a function which does not have an elementary function for its integral. However, there is a trick for getting the total area under the curve. First,

Let,

$$u=\sqrt{\frac{k}{2}}(x-\mu)$$

For getting the total area under the curve. First,

Let

Then,

$$\frac{k}{2}(x-\mu)^2=u^2$$

And,

$$du=\sqrt{\frac{k}{2}}\,dx$$

So,

$$dx=\sqrt{\frac{2}{k}}\,du$$

So our integral becomes,

$$C\int_{-\infty}^{\infty} e^{-\frac{k}{2}(x-\mu)^2}\,dx=C\sqrt{\frac{2}{k}}\int_{-\infty}^{\infty} e^{-u^2}\,du=1$$

Square the integral. When we write down two factors of the integral, we can use different variables for the two integrals.

$$=\frac{2C^2}{k}\left(\int_{-\infty}^{\infty} e^{-x^2}\,dx\right)\left(\int_{-\infty}^{\infty} e^{-y^2}\,dy\right)=1$$

$$= \frac{2C^2}{k} \int_{-\infty}^{\infty} \int_{-\infty}^{\infty} e^{-\left[x^2+y^2\right]} dx\ dy = 1$$

Transfer to polar coordinates,

$$= \frac{2C^2}{k} \int_{0}^{2\pi} \int_{0}^{\infty} re^{-r^2} dr\ d\theta = 1$$

We now have something we can integrate.

Let,

$$v = -r^2$$

$$dv = -2rdr$$

$$rdr = -\frac{1}{2} du$$

So our integral becomes,

$$= \frac{C^2}{k} \int_{0}^{2\pi} \int_{0}^{\infty} e^{-u} du\, d\theta = 1$$

$$= \frac{C^2}{k} \int_{0}^{2\pi} d\theta = 1$$

$$= \frac{2\pi C^2}{k} = 1$$

$$C^2 = \frac{k}{2\pi}$$

or,

$$C = \sqrt{\frac{k}{2\pi}}$$

So,

$$f(x) = \sqrt{\frac{k}{2\pi}} e^{-\frac{k}{2}(x-\mu)^2}$$

Now that we have a formula for the distribution, we can compute its expected value.

Let,

$$x - \mu = v$$

$$dx = dv$$

The expected value of v is,

$$E(v)=\sqrt{\frac{k}{2\pi}}\int_{-\infty}^{\infty} ve^{-\frac{k}{2}v^2}\,dv$$

Let,

$$w=-\frac{k}{2}v^2$$

$$Dw = -\,kvdv$$

$$vdv=-\frac{1}{k}dw$$

So,

$$E(v)=\sqrt{\frac{1}{2\pi k}}\int_{-\infty}^{\infty} e^{w}\quad dw=\sqrt{\frac{1}{2\pi k}}[0-0]=0$$

Hence,

$$E(x)=E(v)+\mu=\mu$$

We can use this to find the variance,

$$\sigma^2=\int_{-\infty}^{\infty}(x-\mu)^2\,f(x)dx$$

$$=\sqrt{\frac{k}{2\pi}}\int_{-\infty}^{\infty}(x-\mu)^2\,e^{-\frac{1}{2}(x-\mu)^2}\,dx$$

Let,

$$W=x-\mu$$

$$dx\ =\ dw$$

And,

$$\sigma^2=\sqrt{\frac{k}{2\pi}}\int_{-\infty}^{\infty} w^2e^{-\frac{k}{2}w^2}\,dw$$

We can use integration by parts to find this.

Let,

$$u = w \quad v = -\frac{1}{k}e^{-\frac{k}{2}w^2}$$

$$du = dw \quad dv = we^{-\frac{k}{2}w^2}\ dw$$

Then our integral becomes,

$$= \sqrt{\frac{k}{2\pi}}\left[-\frac{ve^{-\frac{k}{2}v^2}}{k}\right]_{-\infty}^{\infty} + \frac{1}{k}\sqrt{\frac{k}{2\pi}}\int_{-\infty}^{\infty} e^{-\frac{k}{2}w^2}\ dw$$

The first term is zero and in the second term, after the 1/k, we have the area under the bell shaped curve, with mean = 0, which is 1.

We get,

$$\sigma^2 = \frac{1}{K}$$

Which says,

$$k = \frac{1}{\sigma^2}$$

So the formula becomes,

$$f(x) = \frac{1}{\sigma\sqrt{2\pi}}e^{-\frac{1}{2}\left(\frac{x-\mu}{\sigma}\right)^2}$$

Now that we have the formula, we can locate the critical points in the bell shaped curve. To find maxima and minima, solve,

$$f'(x) = 0$$

$$f'(x) = -\frac{1}{\sigma^3\sqrt{2\pi}}(x-\mu)e^{-\frac{1}{2}\left(\frac{x-\mu}{\sigma}\right)^2} = 0$$

Which is zero if and only if,

$$x - \mu = 0$$

or,

$$x = \mu$$

Which says that the bell shaped curve peaks out above the mean, which we suspected to be true to begin with.

For the points of inflection, solve,

$$f''(x) = 0$$

$$f''(x) = -\frac{1}{\sigma^3\sqrt{2\pi}}\left[1e^{-\frac{1}{2}\left(\frac{x-\mu}{\sigma 1}\right)^2} - \frac{1}{\sigma^2}(x-\mu)^2 e^{-\frac{1}{2}\left(\frac{x-\mu}{\sigma}\right)^2} \right]$$

$$= \frac{1}{\sigma^3\sqrt{2\pi}}\left[\frac{1}{\sigma^2}(x-\mu)^2 - 1\right] e^{-\frac{1}{2}\left(\frac{x-\mu}{\sigma}\right)^2} = 0$$

Which happens if and only if,

$$\left(\frac{x-\mu}{\sigma}\right)^2 - 1 = 0$$

$$\left(\frac{x-\mu}{\sigma}\right)^2 = 1$$

$$\frac{x-\mu}{\sigma} = \pm 1$$

or,

$$x = \mu \pm \sigma$$

This says that the points of inflection in the bell shaped curve lie one standard deviation above and below the mean.

Another advantage of having a formula for the bell shaped curve is that we can use integral calculus to find the areas under portions of the bell shaped curve. The formula does not have an elementary integral even if we allow the use of exponential functions.

To get a method for finding the area under the bell shaped curve, first define is called the z-score,

$$\frac{x-\mu}{\sigma}$$

It tells us how many standard deviations our score lies from the mean,

$$z \equiv \frac{x-\mu}{\sigma}$$

Then,

$$dz = \frac{1}{\sigma}dx$$

or,

$$dx = \sigma dz$$

Then,

$$\frac{1}{\sigma\sqrt{2\pi}}\int e^{-\frac{1}{2}\left(\frac{x-\mu}{\sigma}\right)^2} dx \equiv \frac{1}{\sqrt{2\pi}}\int e^{-\frac{1}{2}z^2} dz$$

Note that,

$$e^{-\frac{1}{2}z^2} = \sum_{n-0}^{\infty}\frac{\left(-\frac{1}{2}z^2\right)^n}{n!}$$

And we can then integrate term by term,

$$\frac{1}{\sqrt{2\pi}}\int e^{-\frac{1}{2}z^2} dz = \frac{1}{\sqrt{2\pi}}\int\sum_{n-0}^{\infty}\frac{\left(-\frac{1}{2}z^2\right)^n}{n!}dz$$

$$= \frac{1}{\sqrt{2\pi}}\sum_{n-0}^{\infty}\int\frac{(-1)^n z^{2n}}{2^n n!}dz$$

$$= \frac{1}{\sqrt{2\pi}}\sum_{n-0}^{\infty}\frac{(-1)^n z^{2n+1}}{2^n(2n+1)n!}$$

This is the area under the bell shaped curve between the mean and our z-score.

It is not that unreasonable that we should need to use an infinite sum to get the area, because one needs to use infinite series to compute ex anyway.

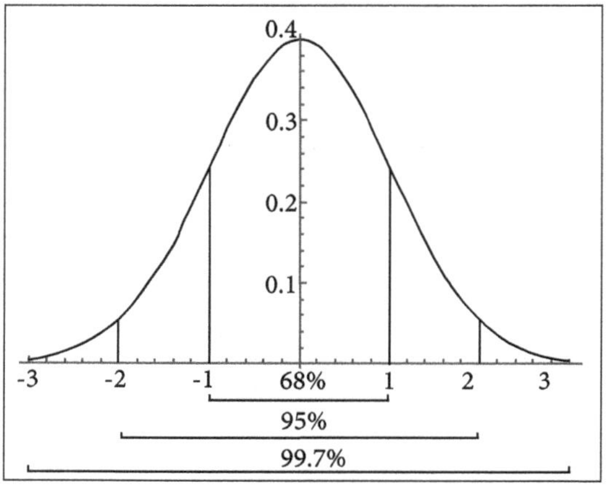

This says that approximately 68% of the scores are within one standard deviation from the mean, 95% of the scores are within two standard deviations from the mean and 99.7% are within 3 standard deviations. Most scores are within standard deviations from the mean.

The Normal Approximation of the Binomial Distribution

In the case of an experiment being repeated n times, if the probability of an event is p, then the probability of the event occurring k times is,

$$ {}_{n}C_{k}\, p^{k} q^{n-k} $$

Where,

$$ q = 1 - p $$

If one were to graph these distributions, it would look somewhat like a bell shaped curve. If the event happening is called a success and it not happening is called a failure, then, if the number of trials is reasonably small, this formula will work fairly well in computing the probabilities of various numbers of successes. However, the probability of a coin landing heads is 1/2 means that if we were to flip a coin a large number of times like, say, 100 times, that heads should come up on half if the tosses or 50 times, using the formula for computing the probability of getting 50 heads in 100 tosses would be quite tedious, but even more tedious would be that, if, after we noticed that this probability of getting exactly 50 heads was rather small, we decided to illustrate the point by computing the probability of getting between 45 and 55 heads, the effort involved in using the binomial formula for getting the probabilities of the eleven outcomes in this event would be somewhat significant. If the binomial probabilities were approximately normally distributed, one could treat it as a bell shaped curve problem.

For a bell shaped curve problem one needs a mean and a standard deviation. It the binomial distribution, it is well known that,

$$\mu = np$$

And,

$$\sigma \equiv \sqrt{npq}$$

The binomial distribution is discrete and the normal distribution is continuous. This will present a challenge in seeing if the binomial distribution satisfies the differential equation which defines the normal distribution.

$$\frac{df}{dx} = -\frac{(x-\mu)}{\sigma^2} f(x)$$

Recall,

$$\frac{df}{dx} = \lim_{h \Rightarrow o} \frac{f(x+h)-f(x)}{h}$$

The smallest value that h could possibly take in this case would be 1.

$$f(x+1)-f(x) = \frac{n(n-1)...(n-x+1)(n-x)}{(x+1)x(x-1)...1} p^{x+1} q^{n-x-1} - \frac{n(n-1)...(n-x-1)}{x(x-1)...1} p^x q^{n-x}$$

$$= \frac{n(n-1)...(n-x-1)}{x(x-1)...1} p^x q^{n-x-1} \left[\frac{n-x}{x+1} p - q\right]$$

$$= \frac{f(x)}{q}\left[\frac{np-xp-xq-q}{x+1}\right]$$

$$= \frac{f(x)}{q}\left[\frac{np-(p+q)x-q}{x+1}\right]$$

But p + q = 1, so we get,

$$= \frac{f(x)}{q}\left[\frac{np-x-q}{x+1}\right]$$

It is time to start approximating. q is a probability, so it is a number between 0 and 1 and as such, it will be negligible compared to np and x, especially when n is large, so we can ignore it. If x + 1 is close to the mean, np, then this becomes close to,

$$\approx \frac{(np-x)f(x)}{npq}$$

$$= -\frac{(x-\mu)f(x)}{\sigma^2}$$

So we see that while the binomial distribution does not give us an exact solution to the differential equation, it is pretty close when x is close to the mean. When x gets farther away from the mean, the probabilities become smaller, so even if the probabilities are off by a significant factor, they are so small that the differences could be negligible compared to the probabilities of outcomes closer to the mean.

In our example of flipping a coin 100 times and noting the number of heads, if one computes the binomial probability of exactly 50 heads, it comes out to 0.079589 (rounded to 6 places after the decimal point), not very much. This is because the probability of 50 heads is not that much different from the probability of 49 or 51 heads or even the probability of 48 or 52 heads. When one spreads the probabilities out over all those possibilities, any one of them will not be that great.

If we figure the binomial probabilities for all of the outcomes between 45 and 55 heads, it comes out to be about 0.728747. If we use the normal distribution, the probability comes of to be about 0.728668. In both cases they round off to 0.7287 agreement to four significant digits, which is not bad, especially for a probability, where that much exactness is not really that meaningful. Notice that even though from 45 to 55 is only around 1/10th of the domain, the probability, when we take these outcomes which are close to half heads and half tails, is over 70%, supporting the contention that we will probably get half heads and half tails.

If on the other hand we try the probability of between 25 and 30 heads, if we use the binomial probabilities, we get around 3.9163×10^{-5}, where if we use the normal distribution we get around 4.7945×10^{-5}. Proportionally, this is a much more significant difference, which is explained by the fact that we are farther from the mean, but either of these numbers and hence their difference, is negligible compared to our previous result.

The upshot is that even though the binomial distribution is not exactly normal, it is close enough that people do use the normal approximation.

Exponential Distribution

A continuous random variable 'X' is said to follow an exponential distribution with parameter $\lambda > 0$ if its probability density function is given by,

$$F(x) = \begin{cases} \lambda e^{-\lambda x} & x > a \\ 0 & \text{otherwise} \end{cases}$$

To find MGF,

$$M_x(t) = \int_{-\infty}^{\infty} e^{tx} f(x) dx$$

$$= \int_{0}^{\infty} e^{tx} \lambda e^{-\lambda x} dx = \lambda \int_{0}^{\infty} e^{-(\lambda - t)x} dx$$

$$= \lambda \left[\frac{e^{-(\lambda-t)x}}{\lambda - t} \right]_0^{\infty}$$

$$= \frac{\lambda}{-(\lambda - t)} \left[e^{-\infty} - e^{-0} \right]$$

$\therefore$ MFG of x $= \dfrac{\lambda}{\lambda - t}, \lambda > t$

To find Mean and Variance:

We know that MGF is given by,

$$M_x(t) = \frac{\lambda}{\lambda - t} = \frac{1}{1 - \frac{t}{\lambda}} = \left(1 - \frac{t}{\lambda}\right)^{-1}$$

$$= 1 + \frac{t}{\lambda} + \frac{t^2}{\lambda^2} + \ldots\ldots + \frac{t^{\tau}}{\lambda^{\tau}}$$

$$= 1 + \frac{t}{\lambda} + \frac{t^2}{2!}\left(\frac{2!}{\lambda^2}\right) + \ldots\ldots + \frac{t^{\tau}}{r!}\left(\frac{t!}{\lambda^{\tau}}\right)$$

$$M_x(t) = \sum_{\tau - 0}^{\infty} \left(\frac{t}{\lambda}\right)^{\tau}$$

$$\therefore \text{Mean} \quad \mu_1' = \text{Coefficient of } \frac{t^1}{1!} = \frac{1}{\lambda}$$

$$\mu_2' = \text{Coefficient of } \frac{t^2}{2!} = \frac{2}{\lambda^2}$$

$$\text{Varriance} = \mu_2 = \mu_2' - \mu_1'^2 = \frac{2}{\lambda^2} - \frac{1}{\lambda^2} = \frac{1}{\lambda^2}$$

$$\text{Varrriance} = \frac{1}{\lambda^2} \qquad \text{Mean} = \frac{1}{\lambda}$$

Problems

1. If X and Y are independent exponential random variables each with parameter 1, let us determine the p.d.f of U = X - Y.

Solution:

Given:

$$U = X - Y$$

p.d.f. of U=X-Y

$$x = u + v, \ y = v \quad J = \begin{vmatrix} \frac{\partial x}{\partial u} & \frac{\partial x}{\partial u} \\ \frac{\partial y}{\partial u} & \frac{\partial y}{\partial V} \end{vmatrix} = 1$$

The joint p.d.f. of (u, v) is f(u, v) = |J| f(x, y) = $e-^{(u + 2v)}$

The range space of (u, v) is V - u when a o and v o when u o.

$$f_u(u) = \int_0^{\infty} e^{-(u+2v)} dv \text{ when o}$$

$$= \int_0^{\infty} e^{(-u+2v)} dv \text{ when o}$$

$$f_u(u) = \frac{1}{2} e^{-u}, u < o$$

$$= \frac{1}{2} e^{-u}, u > o.$$

2. Let 'X' be a random variable with p.d.f $F(x) = \begin{cases} \frac{1}{3} e^{\frac{-x}{3}} & x > o \\ o & \text{otherwise} \end{cases}$

Then let us determine:

- P(X > 3)
- MGF of 'X'.

Solution:

Given:

$$F(x) = \begin{cases} \frac{1}{3} e^{\frac{-x}{3}} & x > o \\ o & \text{otherwise} \end{cases}$$

Formula to be used:

WKT the exponential distribution is given by,

$$F(x) = \lambda e^{-\lambda x}, \quad x > 0$$

Here,

$$\lambda = \frac{1}{3}$$

$$P(x > 3) = \int_{3}^{\infty} f(x)dx \qquad = \int_{3}^{\infty} \frac{1}{3} e^{\frac{x}{3}} dx$$

$$P(X > 3) = e^{-1}$$

MGF is $M_x(t) = \dfrac{\lambda}{\lambda - 1}$

$$= \frac{\frac{1}{3}}{\frac{1}{3} - t} = \frac{\frac{1}{3}}{\frac{1-3t}{3}} = \frac{1}{1-3t}$$

$$M_X(t) = \frac{1}{1-3t}$$

Note:

If X is exponentially distributed, then $P(X > s + t/x > s) = P(X > t)$, for any $s, t > 0$.

1.3.1 Distribution of Several Random Variables

We only consider the case of two continuous variables. The extension to more variables is straightforward. The infinitesimal element of probability is $f(x,y) = \dfrac{\partial^2 F(x,y)}{\partial x \partial y}$ and the probability density function,

$$\iint_A f(x,y)dx\,dy. \quad ...(1)$$

The probability of finding the variable inside a certain area $P(|X| > 3) = 0.27\%$ is

$$f_X(x) \qquad ...(2)$$

Marginal Distributions

$$\int_{-\infty}^{+\infty} f(x,y)dy, \quad [\bar{X}_{300}] \quad f_Y(y) \quad ...(3)$$

$$\int_{-\infty}^{+\infty} f(x,y)dx, \quad [\bar{X}_{300}] \quad [XY]- \quad ...(4)$$

The subscripts,

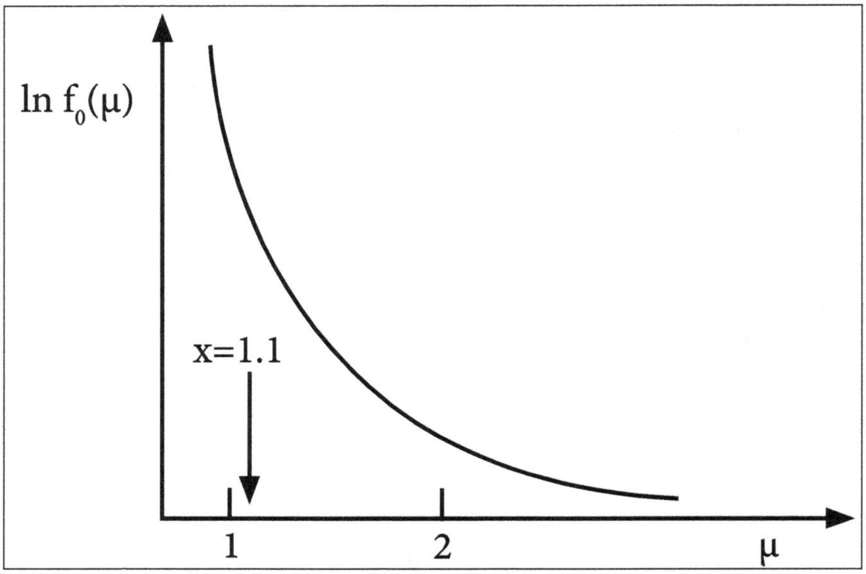

And Xi indicate that $f_Y(y)$ and $f_X(x/y)$ are only functions of,

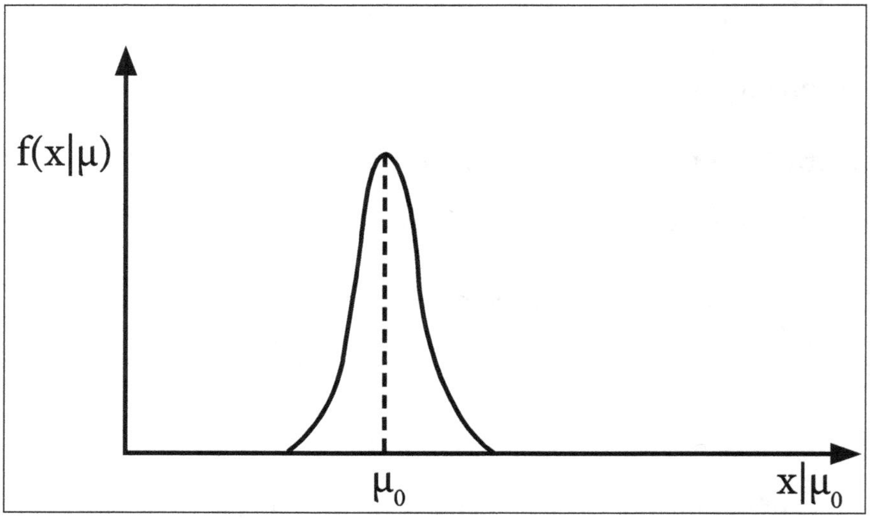

And X_i, respectively, but in most cases we will drop the subscripts if the context helps in resolving ambiguities.

Conditional Distributions

$$\frac{f(x,y)}{f_Y(y)} = \frac{f(x,y)}{\int f(x,y)dx}, [\bar{X}_{300}] f_Y(y|x) \quad ...(5)$$

$$\frac{f(x,y)}{f_X(x)}, \quad [\bar{X}_{300}] \quad f(x,y) \quad ...(6)$$

$$f_X(x|y)\, f_Y(y) [\bar{X}_{300}] f_Y(y|x) f_Y(x) \quad ...(7)$$

$$[\bar{X}_{300}] f(x,y) = f_X(x) f_Y(y) \quad ...(8)$$

Independent random variables,

$$f_X(x/y) = f_X(x) \qquad ...(9)$$

(It implies $f_Y(y/x) = f_Y(y)$ and $f(h|e) = \frac{f(e|h) f_h(h)}{\int f(e|h) f_h(h) dh}$)

Bayes' theorem for continuous random variables,

$$\mu X = \quad ...(10)$$

Expectation Value

$$\iint_{-\infty}^{+\infty} x\, f(x,y) dx\, dy \quad {}_E\int_{-\infty}^{+\infty} x\, f(x) dx, \quad [\bar{X}_{300}] \quad \int_{-\infty}^{+\infty} x\, fx(x) dx, \qquad ...(11)$$

$$[\bar{X}_{300}] \quad [g(X,Y)] = \int\int_{-\infty}^{+\infty} g(x,y) f(x,y) dx\, dy \qquad ...(12)$$

And analogously for X_i. In general

$$E\sigma^2 x = \qquad ...(13)$$

Variance

$$[X^2] - E^2[X], E(X,Y) \qquad ...(14)$$

And analogously for xi.

Covariance

$$[(X - E|X|)\ (Y - E|Y|)] [\bar{X}_{300}] E[XY] - \quad ...(15)$$

$$\left[X_{300}\right] E[X]E\left[Y\right] E[XY] \qquad ...(16)$$

If,

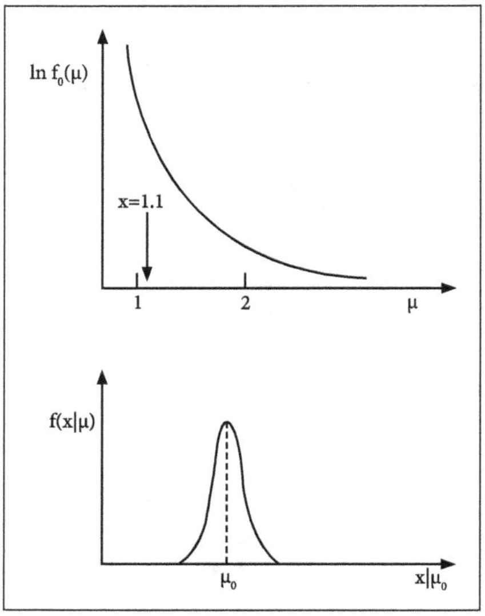

And Xi are independent, then E[X]E[Y]E(X, Y)=0 and hence Cov Y ~ N(.) (the opposite is true only if Y ~ N(.)).

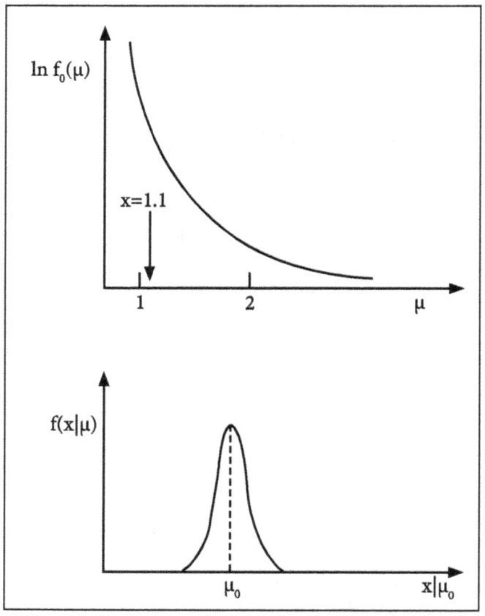

Correlation Coefficient

$$\frac{\mathrm{Cov}(X,Y)}{\sqrt{\mathrm{Var}(X)\mathrm{Var}(Y)}} \left[\bar{X}_{300}\right] \frac{\mathrm{Cov}(X,Y)}{\sigma_X \sigma_Y}$$

$$\left[\bar{X}_{300}\right](-1 \leq \rho \leq 1)$$

$$Y = \sum_{i} C_i X_i$$

Linear Combinations of Random Variables

If ci, with μY= real, then,

$$[Y]E\sum_{i} c_i \left[\bar{X}_{300}\right](X_i) = \sum_{i} c_i \mu_i, E\sigma_Y^2 =$$

$$(Y)\text{Var}\sum_{i} c_i^2 \left[\bar{X}_{300}\right](X_i) + 2\sum_{i<j} c_i c_j \text{ Var}(X_i X_j)\text{Cov}(X_i) + \sum_{i \neq j} c_i c_j$$

$$\left[\bar{X}_{300}\right](X_i) + 2\sum_{i<j} c_i c_j \text{ Var}\sum_{i} c_i^2 \sigma_i^2 + \sum_{i \neq j} \rho_{ij} c_i c_j \sigma_i \sigma_j \text{Cov}(X_i) + \sum_{i \neq j} c_i c_j$$

$$\left[\bar{X}_{300}\right] \sum_{ij} \rho_{ij} c_i c_j \sigma_i \sigma_j$$

$$\left[\bar{X}_{300}\right] \sum_{ij} c_i c_j \sigma_{ij}$$

$$\left[\bar{X}_{300}\right] \sigma_Y^2$$

σ_{ij} has been written in different ways, with increasing levels of compactness, that can be found in the literature.

In particular (X_i) and (X_j) and use the notations $(X_i, X_j) = \rho_{ij}\sigma_j\sigma_i$ Cov $\sigma_{ii} = \sigma_{i2}$ and $\rho_{ii} = 1$ and the fact that, by definition, ρ.

Bivariate normal distribution: Joint probability density function of,

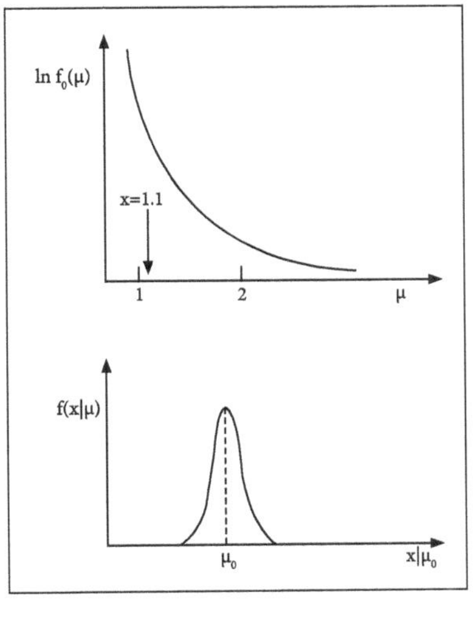

And Xi with correlation coefficient,

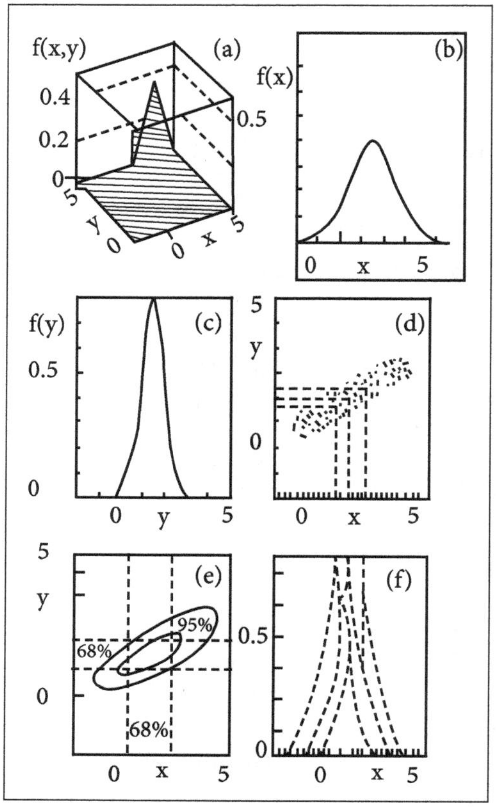

Example of bivariate normal distribution.

$$\frac{1}{2\pi\sigma_x\sigma_y\sqrt{1-\rho^2}}.$$

$$f_X(x\,|\,y)f_Y(y) \quad \left[\bar{X}_{300}\right] \quad \exp\left\{-\frac{1}{2\left(1-\rho^2\right)}\left[\frac{\left(x-\mu_x\right)^2}{\sigma_x^2}-2\rho\frac{\left(x-\mu_x\right)\left(y-\mu_y\right)}{\sigma_x\sigma_y}+\frac{\left(y-\mu_y\right)^2}{\sigma_y^2}\right]\right\}$$

Marginal Distributions

$X \sim N\left(\mu x,\ \sigma x\right)$, Y

$$N\left(\mu_y,\sigma_y\right) \quad N\left(\mu_x,\sigma_x\right), \quad f\left(y\,|\,x_o\right)=\frac{1}{\sqrt{2\pi}\,\sigma_y\sqrt{1-\rho^2}}\exp\left[-\frac{\left(y-\left[\mu_y+\rho\frac{\sigma_y}{\sigma_x}\left(x_o-\mu_x\right)\right]\right)^2}{2\sigma_y^2\left(1-\rho^2\right)}\right],$$

Conditional Distribution

$$Y_{|xo} \sim N\left(\mu_y + \rho\frac{\sigma_y}{\sigma_x}(x_o - \mu_x), \sigma_y\sqrt{1-\rho^2}\right).$$

i.e.,

$X = x_0$ The condition $Y = \sum_{i=1}^{n} c_i X_i$, squeezes the standard deviation and shifts the mean of X_i .

1.4 Statistics: Random Sampling

Random Sampling

It is often necessary to prove that the data set is indeeds randomly chosen part of the population. If the sample produces unpopular results end conclusions, the most common defence is an attack as to the randomness of the sample.

Example: An auditor has a population of 1000 invoices which he is responsible for checking. He hes decided that is 5% check is sufficient. In other words he intends going through only 50 invoices thoroughly. If his sample indicates a reasonable level of accuracy he will then assume that the entire population is also reasonably accurate. To ensure that his sample is truly random he decides en use a random number table. Each invoice is numbered sequentially from 1 to 1000. His sample of 50 could be generated as follows:

- Without looking, he places his finger somewhere on the random number table. His finger seems to be pointing at the number 6312 which occurs in the 16th row and the 6th column. Now 6312 obviously exceeds 1000 so it doesn't correspond to any invoice number. Therefore he drops the lest digit, the 2. He is left with the number 631, corresponding to invoice number 631. This is the first invoice that has to be checked.
- Immediately below 6312 lies the number 4 003. Again he drops the lest digit, a 3 and invoice number 400 becomes the second invoice to be placed in his random sample.
- Immediately below 4003 lies the number 3599. Again he drops the last digit, 9 and invoice number 359 becomes the third invoice to he placed in his random sample.
- In this fashion, he continues down the sixth column, choosing an invoice corresponding. every number scrutinized. He then starts down the seventh column, only stepping after the 50th random number. Duplicate random numbers

may mean that one invoice is included in the sample more than once. This does not matter. Note that the random number 000 has to be used to denote invoice number 1000 since only three digit numbers are considered by this method.

- If an estimate for the mean invoice amount is also required, the total amount for each of the 50 invoices selected has to be recorded. The average of the 50 invoice totals is then an estimate of the average of the 1990 invoice totals from the population of 1000 invoices.
- If estimate what proportion of the 1 000 invoices contains errors, the auditor can find out (by checking each of the 60 invoices in the sample) what proportion of the sample of 50 invoices contains errors.

Of course there are many other ways of generating a random sample using a random number table. One can read across rather than down. One can read only every second number, etc. As long as one is systematic in that the same rule is always used to obtain the next invoice. The fact that the sample of invoices chosen for checking is truly random. Hopefully the proportion of invoices in error in the sample of 50 will be close to the proportion of invoices in error in the population of 1000 invoices. This may not be the ease: it is conceivable, although highly unlikely, that the auditor will pick out the only 50 invoices in the entire population that contain errors. But this is another problem which will be considered at a later stage. At least the auditor has ensured that he cannot be accused of having allowed personal bias to affect his choice of sample.

1.4.1 Estimation of Parameters and Confidence Intervals

Parameter Estimation

There are number of population parameters of potential interest when one is estimating health outcomes (or "endpoints"). Many of outcomes we are interested in estimating are either continuous or dichotomous variables. The parameters to be estimated depend not only on whether the endpoint is continuous or dichotomous, but also on the number of groups being studied. Moreover, when two groups are being compared, it is important to establish whether the groups are independent (e.g., men versus women) or dependent (i.e., matched or paired, such as a before and after comparison). The table below summarizes parameters that may be important to estimate in health-related studies.

	Parameters Being Estimated	
No. of Sample	Continuous Variable	Dichotomous Variable
One Sample	Mean	proportion or rate, e.g., prevalence, cumulative incidence, incidence rate
Two Independent Samples	difference in means	difference in proportions or rates, e.g., risk difference, rate difference, risk ratio, odds ratio, attributable proportion

Two Dependent, Matched Samples	mean difference	

Confidence Intervals

There are 2 types of estimates for each population parameter: the point estimate and confidence interval (CI) estimate. For both continuous variables (e.g., population mean) and dichotomous variables (e.g., population proportion) one first computes the point estimate from a sample. Recall that sample means and sample proportions are unbiased estimates of the corresponding population parameters.

For both continuous and dichotomous variables, the confidence interval estimate (CI) is a range of likely values for the population parameter based on:

- The point estimate, e.g., the sample mean.
- The investigator's desired level of confidence (most commonly 95%, but any level between 0-100% can be selected).
- The sampling variability or the standard error of the point estimate.

Strictly speaking a 95% confidence interval means that if we were to take 100 different samples and compute a 95% confidence interval for each sample, then approximately 95 of the 100 confidence intervals will contain the true mean value (μ). In practice, however, we select one random sample and generate one confidence interval, which may or may not contain the true mean. The observed interval may over- or underestimate μ. Consequently, 95% CI is the likely range of the true, unknown parameter. The confidence interval does not reflect the variability in the unknown parameter. Rather, it reflects the amount of random error in the sample and provides a range of values that are likely to include the unknown parameter. Another way of thinking about a confidence interval is that it is the range of likely values of the parameter (defined as the point estimate + margin of error) with a specified level of confidence (which is similar to a probability).

Suppose we want to generate a 95% confidence interval estimate for an unknown population mean. This means that there is a 95% probability that the confidence interval will contain the true population mean. Thus, P ([sample mean] - margin of error < μ < [sample mean] + margin of error) = 0.95.

The Central Limit Theorem introduced in the module on Probability stated that, for large samples, the distribution of the sample means is approximately normally distributed with a mean,

$$\mu_{\bar{x}} = \mu$$

And a standard deviation (also called the standard error),

$$\sigma_{\bar{x}} = \sigma/\sqrt{n}$$

For the standard normal distribution, $P(-1.96 < Z < 1.96) = 0.95$, i.e., there is a 95% probability that a standard normal variable, Z, will fall between -1.96 and 1.96. The Central Limit Theorem states that for large samples,

$$z = \frac{\bar{X} - \mu_{\bar{x}}}{\sigma_{\bar{x}}} = \frac{\bar{X} - \mu}{\sigma/\sqrt{n}}$$

By substituting the expression on the right side of the equation,

$$P\left(-1.96 < \frac{\bar{X} - \mu}{\sigma/\sqrt{n}} < 1.96\right) = 0.95$$

Using algebra, we can rework this inequality such that the mean (μ) is the middle term, as shown below,

$$P\left(-1.96\sigma/\sqrt{n} < \mu < \bar{X} + 1.96\sigma/\sqrt{n}\right) = 0.95$$

Then,

$$P\left(-1.96\sigma/\sqrt{n} < \mu < \bar{X} + 1.96\sigma/\sqrt{n}\right) = 0.95$$

And finally,

$$P\left(-1.96\sigma/\sqrt{n} < \mu < \bar{X} + 1.96\sigma/\sqrt{n}\right) = 0.95$$

This last expression, then, provides the 95% confidence interval for the population mean and this can also be expressed as,

$$\bar{X} \pm 1.96\sigma/\sqrt{n}$$

Thus, the margin of error is 1.96 times the standard error (the standard deviation of the point estimate from the sample) and 1.96 reflects the fact that a 95% confidence level was selected. So, the general form of a confidence interval is:

Point Estimate + Z SE (Point Estimate)

Where Z is the value from the standard normal distribution for the selected confidence level (e.g., for a 95% confidence level, Z=1.96).

In practice, we often do not know the value of the population standard deviation (σ). However, if the sample size is large (n > 30), then the sample standard deviations can be used to estimate the population standard deviation.

Z-Scores for Commonly Used Confidence Intervals:

Desired Confidence Interval	Z Score
90%	1.645
95%	1.96
99%	2.576

In the health-related publications a 95% confidence interval is most often used, but this is an arbitrary value and other confidence levels can be selected. Note that for a given sample, the 99% confidence interval would be wider than the 95% confidence interval, because it allows one to be more confident that the unknown population parameter is contained within the interval.

1.5 Testing of Hypothesis and Acceptance Sampling

There are two types of samples in testing of hypothesis:

- Small Sample (n <30).
- Large Sample (n >30).

Small Sample: (n <30)

Type 1 Student's 't' test:

- Single Mean (S.D given directly).
- Single Mean (S.D is not given directly).
- Difference of Means.

Large Sample: (n >30)

- Type 1 Single Mean.
- Type 2 Single Proportion.
- Type 3 Difference of Mean.
- Type 4 Difference of Proportion.

Test of Significance for Large Samples

For large n (sample size), almost all the distributions can be approximated very closely by a normal probability curve, we therefore use the normal test of significance for large samples,

$$Z = \frac{t - E(t)}{\sqrt{V(t)}} = N(0.1)$$

Thus if the discrepancy between the observed and the expected (hypothetical) value of a statistics is greater than Zα times the standard error (S.E), hypothesis is rejected at α level of significance. Similarly to give by,

$$t\ E(t) \leq Z\alpha\ S.E\ (t),$$

The deviation is not regarded significant at 5% level of significance. In other words the deviation.

t- E(t), could have arisen due to fluctuations of sampling and the data do not provide any evidence against the null hypothesis which may, therefore be accepted at α level of significance.

If $Z \leq 1.96$, then the hypothesis Ho is accepted at 5% level of significance.

Thus the steps to be used in the normal test are as follows:

- Compute the test statistic Z under H_o.
- If $Z > 3$, H_o is always rejected.
- If $Z < 3$, we test its significance at certain level of significance.

Test for Single Mean

The tests of significance for variables is that the sample mean is asymptotically normally distributed even if the parent population from which the sample is drawn is not normal.

If xi (i = 1,n) is a random sample of size n from a normal population with mean and variance σ2, then the sample mean is distributed normally with mean and variance $\frac{\sigma^2}{n}$,

$$\bar{x} \sim N\left(\mu_2 \frac{\sigma^2}{N}\right)$$

H_0: There is no significant difference between sample mean $(\overline{X})$ and population mean.

Test Statistic,

$$Z = \frac{\overline{x} - \mu}{\sigma/\sqrt{n}} \sim N(0,1)$$

If σ_2 is unknown, then it is estimated by sample variance i.e., $\sigma^2 = s^2$ (for large n).

Test for Difference of Means

Let $\overline{x}_1 (\overline{x}_2)$ be the mean of a sample of size n_1 (n_2) from a population with mean μ_1 (μ_1) and variance $\sigma_1^{\,2}$ $(\sigma^2{}_2)$. Therefore,

$$\overline{x}_1 \sim N\left(\mu_1, \frac{\sigma_1^2}{n_1}\right), \quad \overline{x}_2 \sim N\left(\mu_2, \frac{\sigma_2^2}{n_2}\right)$$

The different $(\overline{x_1} - \overline{x_2})$ is also normal variate test statistic:

$$Z = \frac{\overline{x}_1 - \overline{x}_2 - E(\overline{x}_1 - \overline{x}_2)}{S.E(\overline{x}_1 - \overline{x}_2)} \sim N(0,1)$$

$$Z = \frac{\overline{x}_1 - \overline{x}_2 - (\mu_1 - \mu_2)}{\sqrt{\frac{\sigma_1^2}{n_1} + \frac{\sigma_2^2}{n_2}}}$$

Under the null hypothesis = $H_0 : \mu_1 = \mu_2$

$$Z = \frac{(\overline{x}_1 - \overline{x}_2)}{\sqrt{\frac{\sigma_1^2}{n_1} + \frac{\sigma_2^2}{n_2}}} \sim N(0,1)$$

Therefore,

$$Z = \frac{(\overline{x}_1 - \overline{x}_2)}{\sigma\sqrt{\frac{1}{n_1} + \frac{1}{n_2}}}$$

If $\sigma_1^2 = \sigma_2^2 = \sigma^2$

If σ is not known, then its estimate is used.

$$\hat{\sigma}^2 = \frac{n_1 s_1^2 + n_2 s_2^2}{n_1 + n_2}$$

Hypothesis testing applications with a continuous outcome variable in a single population are performed according to the five-step procedure outlined above. A key component is setting up the null and research hypotheses. The objective is to compare the mean in a single population to known mean (μ_o).

The known value is generally derived from another study or report, for example a study in a similar, but not identical, population or a study performed some years ago. The latter is known as historical control. It is necessary in setting up the hypotheses in a one sample test which the mean specified in the null hypothesis is a fair and reasonable comparator.

In one sample tests for a continuous outcome, we set up our hypotheses against an appropriate comparator. We select a sample and compute descriptive statistics on the sample data - including the sample size (n), the sample mean $(\bar{X})$ and the sample standard deviation (s). We then determine the appropriate test statistic for the hypothesis test. The formulas for test statistics depend on the sample size and are given below:

Test Statistics for Testing $H_o: \mu = \mu_o$	
if $n \geq 30$	if $n < 30$
$Z = \frac{\bar{X} - \mu_o}{s / \sqrt{n}}$	$t = \frac{\bar{X} - \mu_o}{s / \sqrt{n}}$ Where df=n-1

Procedure involved in testing the significance for difference of proportions in the case of large samples.

The proportional to successes in two large samples of size n_1 and n_2 respectively drawn from the same population or from two populations with the same proportion P. Then

p_1 follows $N\left[P, \sqrt{\frac{PQ}{n_1}}\right]$ p_2 and follows $N\left[P, \sqrt{\frac{PQ}{n_2}}\right]$. Therefore $p_1 - p_2$, which is a linear combination of two normal variables, also follows a normal distribution.

$$E[p_1 - p_2] = E[p_1] - E[p_2] = P - P = 0$$

$$V[P_1 - P_2] = V(p_1) + V(p_2)$$

$$= PQ\left(\frac{1}{n_1}+\frac{1}{n_2}\right)$$

Now,

($\because$ two samples are independent)

$\therefore (p_1 - p_2)$ follows $N\left[o, \sqrt{PQ\left(\frac{1}{n_1}+\frac{1}{n_2}\right)}\right]$

$\therefore$ The test statistic,

$$z = \frac{p_1 - p_2}{\sqrt{PQ\left(\frac{1}{n_1 + \frac{1}{n_2}}\right)}}$$

If P is not known, an unbiased estimate of P based on both samples, given by $\hat{P} = \frac{n_1 p_1 + n_2 p_2}{n_1 + n_2}$, is used in the place of P.

If $|Z| \le Z_\alpha$, then the difference between the two sample proportions p1 and p2 is not significant at α% LOS.

Methods Involved in Testing of Hypothesis

- Set up the null hypothesis.
- Choose the appropriate level of significance which is to be decided before sample is drawn.
- Compute the test statistic $Z = \frac{t - E(t)}{sE(t)}$ under the null hypothesis.
- We compare the computed value of z in step (iii) with the significant value at given level of significance:
    - If |z| < 1.96, Ho may be accepted at 5% level of significance.
    - If |z| > 1.96, Ho may be rejected at 5% level of significance.
    - If |z| < 2.58, Ho may be accepted at 1% level of significance.
    - If |z| > 2.58, Ho may be rejected at 1% level of significance.

Null and Alternate Hypothesis

Null Hypothesis

For applying the test of significance, we first set up a hypothesis which is a statement about the population parameter. This statement is usually a hypothesis of no difference and it is denoted by H_o.

Alternative Hypothesis

If the null hypothesis is false, then something else must be true. This is called an alternate hypothesis and it is denoted by H_1.

Procedure followed in testing of a hypothesis:

- Null Hypothesis H_o is defined.
- Alternative Hypothesis H_1 is also defined after a careful study of the problem.
- The nature of the test (whether one-failed or two failed) is decided.
- Level of Significance LOS 'α' is fixed or taken from the problem of specified and Z_α is noted.
- The test-statistic $Z = \frac{t - E(t)}{S.E(t)}$ is computed.
- Comparison is made between |Z| and Z_α:
    - If $|Z| < Z_\alpha$, H_o is accepted or H1 is rejected i.e., the difference between t and E(t) is not significant at α% LOS.
    - If $|Z| > Z_\alpha$, H_o is rejected or H1 is accepted i.e., the difference between t and E(t) is significant at α% LOS.

Chi Square Test for Goodness of Fit

A test of wide applicability to numerous problems of significance in frequency data is the χ^2 test of goodness of fit. It is primarily used for testing the discrepancy between the expected and the observed frequency, for instance, in comparing an observed frequency distribution with a theoretical one like the normal.

H_o: The fitted distribution is a good fit to the given data.

H_1: Not a good fit.

Test Statistic

If O_i and E_i, i = 1,...,n are the observed and expected frequency of ith class respectively, then the statistic,

$$\Psi^2 = \sum \frac{(O-E)^2}{E}$$

Where,

r is the number of parameters estimated from the sample.

n is the number of classes after pooling.

H_0 is rejected at level α if calculated χ^2 > tabulated $_{n-r-1}(\alpha)$.

Problems

1. An automobile company gives us the following information about age groups and the linking for particular model of car which it planes to introduce. On the basis of this data, let us check if it can be concluded that the model appeal is independent of the age group $\left(\chi^2 0.05\ (3)\ =\ 7.815\right)$.

Solution:

Given:

Age					
Persons	Below 20	20-39	40-59	60>	total
Liked the car	140	80	40	20	280
Disliked the car	60	50	30	80	220
Total	200	130	70	100	500

Table of expected frequencies:

Below 20	20-39	40-59	60>	total
$\frac{200\times280}{500}=112$	$\frac{130\times280}{500}=72.8$	$\frac{70\times280}{500}=39.2$	$\frac{100\times280}{500}=56$	280
$\frac{200\times220}{150}=88$	$\frac{130\times220}{500}57.2$	$\frac{70\times220}{500}=30.8$	$\frac{100\times220}{500}=44$	220
222	130	70	100	500

Calculation of Ψ^2

O	E	$(O-E)^2$	$(O-E)^2/E$
140	112	28^2	7

80	72.8	$(7.2)^2$	0.712
40	39.2	$(0.8)^2$	0.016
20	56	$(-36)^2$	23.14
60	88	$(-28)^2$	8.909
50	57.2	$(-7.2)^2$	0.906
30	30.8	$(-0.8)^2$	0.02
80	44	$(36)^2$	29.45
Total			70.154

$\psi^2 = 70.154$

Degree of freedom, V = (4-1) (2-1)

= 3

$t^3_{0.05} = 7.815$

Calculated value > table value

∴ H_0 is rejected.

2. Let us fit a Poisson distribution to the following data and test the goodness of fit.

X:	0	1	2	3	4	5
Frequency f(x):	275	72	30	7	5	2

Solution:

Given:

X:	0	1	2	3	4	5
Frequency f(x):	275	72	30	7	5	2

To find the Poisson distribution whose probability law is,

$$P(X = x) = \frac{e^{-\lambda}\lambda^x}{x!}; x = 0,1,2,.....$$

We need λ (Mean of Poisson distribution).

x:	0	1	2	3	4	5	6	Total
f:	275	72	30	7	5	2	1	392
f.x:	0	72	60	21	20	10	6	189

$$\lambda = \frac{\sum fx}{\sum f} = \frac{392}{189} = 2.07$$

The expected frequencies are given by,

$$\frac{N \cdot e^{-\lambda} \lambda^x}{x!} = \frac{(392) e^{-2.07} (2.07)^x}{x!} ; x = 0,1,2,....,6.$$

x:	0	1	2	3	4	5	6
Ei:	49.465	102.392	105.976	73.124	37.841	15.66	5.405
Rounded:	49	102	106	73	38	16	5
Oi:	275	72	30	7	5	2	1

Combining the end classes so as to make the individual frequencies greater than 10.

Oi	Ei	$(O_i - E_i)^2 / E_i$
275	49	1042.37
72	102	8.82
30	106	54.49
15	132	103.70
Total		$\chi^2 = 1209.38$

H_0: The given distribution is nearly Poisson distribution.

Degree of freedom, V = n − 2 = 4 − 2 = 2.

$$\chi^2 \; 0.05(V = 2) = 5.99$$

Here calculated x^2 > Tabulated

∴ H_0 is rejected.

Problems

1. The heights of college students in Chennai are normally distributed with the standard deviation 6cm and sample of 100 students had their mean height 158cm. Let us test the hypothesis that the mean height of college students in Chennai is 160cm at 1% level of significance.

Solution:

Given:

$$n = 100$$

$\overline{x} = 158$

$\mu = 160$

$\sigma = 6$

$H_0 : \overline{x} = \mu$ (The difference between x and μ is not significant)

$H_1 : \overline{x} \neq \mu$

$$z = \frac{\overline{x} - \mu}{\sigma/\sqrt{n}} = \frac{158 - 160}{6/10} = -3.33$$

$|Z| = 3.33.$

We know that,

$Z_{0.01} = 2.58$; Here $|Z| > Z_{0.01}$

∴ H_0 is rejected.

∴ The sample is not drawn from the population with mean height, μ = 160 cm.

2. A machine produces 16 imperfect articles in a sample of 500. After machine is overhauled, it produces 3 imperfect articles in a batch of 100. Let us check whether the machine has been improved or not.

Solution:

Given:

$n_1 = 500$

$$p_1 = \frac{16}{500} = 0.032$$

$n_2 = 100$

$$p_2 = \frac{3}{100} = 0.03$$

$H_0 : P_1 = P_2 = P$

$H_1 : P_1 \neq P_2$

The test statistic is,

$$Z = \frac{p_1 - p_2}{\sqrt{pq\left(\frac{1}{n_1} + \frac{1}{n_2}\right)}}$$

where,

$$p = \frac{n_1 p_1 + n_2 p_2}{n_1 + n_2}$$

$$= \frac{500(0.032) + 100(0.03)}{500 + 100} = \frac{19}{600}$$

$$= 0.0316$$

$$q = 1 - 0.0316$$

$$= 0.9684$$

$$\therefore Z = \frac{0.032 - 0.03}{\sqrt{(0.0316)(0.9684)\left(\frac{1}{500} + \frac{1}{100}\right)}}$$

$$= \frac{0.002}{\sqrt{(0.0307)(0.012)}} = \frac{0.002}{0.0191}$$

$$|Z| = 0.104$$

Since, $|Z| < 1.96$. We accept the null hypothesis at the level of significance.

3. A manufacturer of bulbs claims that an average of 2% of the bulbs manufactured by him are defective. A random sample of 400 bulbs contained 13 defective bulbs. On the basis of the sample, let us check that we can support the manufacturers claim at 5% level of significance.

Solution:

Given:

$$n = 400$$

$$p = \frac{13}{400} = 0.0325$$

Sample proportion,

Population proportion,

$$P = 2\% = 0.02.$$

$$\therefore\ Q = 1 - P = 1 - 0.02 = 0.98.$$

Null HypothesisH_0: P = 0.02

Alternate Hypothesis H_1: P > 0.02 (one tail test)

The test statistic,

$$z = \frac{p - P}{\sqrt{\frac{PQ}{n}}}$$

$$z = \frac{0.0325 - 0.02}{\sqrt{\frac{(0.02)(0.98)}{400}}} = 1.7857$$

$$|z| = 1.7857$$

At 5% level of significance for one tail test, $z_{\alpha} = 1.645$.

Here, calculated z value = 1.7857 is greater than tabulated z_{α} Value = 1.645.

Hence, we can reject the null hypothesis.

Example: The manufactures cannot be supported.

1.6 Regression Analysis

Regression analysis is a widely used technique which is useful for evaluating multiple independent variables. As a result, it is particularly useful for assess and adjusting for confounding. It can also be used to assess the presence of effect modification.

Simple Linear Regression

Suppose we want to assess the association between the total cholesterol and body mass index (BMI) in which the total cholesterol is the dependent variable and BMI is the independent variable. In regression analysis, the dependent variable is denoted Y and the independent variable is denoted X. So, in this case, Y=total cholesterol and X=BMI.

When there is a single continuous dependent variable and a single independent variable, the analysis is called as simple linear regression analysis. This analysis assumes that there is a linear association between the two variables. (If a different relationship is hypothesized, such as a curvilinear or exponential relationship, the alternative regression analyses are performed).

The figure below is a scatter diagram illustrating the relationship between the BMI and total cholesterol. Each point represents the (X, Y) pair, in this case, BMI and the corresponding total cholesterol measured in each participant. Note that the independent variable is on the horizontal axis and the dependent variable on the vertical axis.

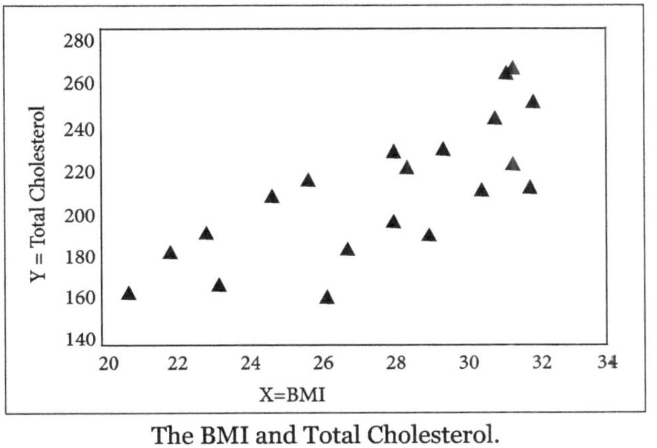

The BMI and Total Cholesterol.

The graph shows that there is a positive or direct association between the BMI and total cholesterol, participants with lower BMI are more likely to have lower total cholesterol levels and participants with higher BMI are more likely to have higher total cholesterol levels. In contrast, suppose we examine the association between BMI and HDL cholesterol.

In contrast, the graph below depicts the relationship between BMI and HDL HDL cholesterol in the same sample of n=20 participants.

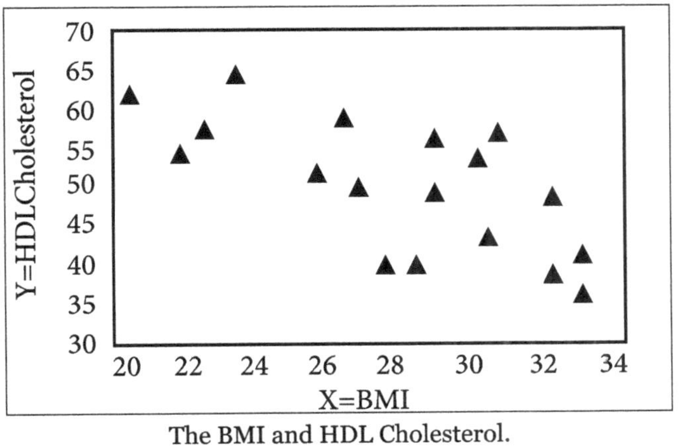

The BMI and HDL Cholesterol.

This graph shows a negative or inverse association between the BMI and HDL cholesterol, i.e., those with lower BMI are more likely to have higher HDL cholesterol levels and those with higher BMI are more likely to have lower HDL cholesterol levels.

For either of these relationships we could use simple linear regression analysis to estimate the equation of the line that best describes the association between the independent variable and the dependent variable. The simple linear regression equation is as follows,

$$\hat{Y} = b_o + b_1 X$$

Where $\hat{Y}$ is the predicted or expected value of the outcome, X is the predictor, bo is the estimated Y-intercept and b1 is the estimated slope. Y-intercept and slope are estimated from the sample data so as to minimize the sum of the squared differences between the observed and the predicted values of the outcome, i.e., the estimates minimize,

$$\sum\left(Y-\hat{Y}\right)^2$$

These differences between observed and predicted values of the outcome are called residuals. The estimates of the Y-intercept and slope minimize the sum of the squared residuals and are called the least squares estimates.

1.6.1 Fitting Straight Lines

Given a set of points,

$$(x_1, y_1), \ldots, (x_n, y_n)$$

How do we find a straight line,

$$y = a + bx$$

Which provides a good description of the general trend underlying the points?

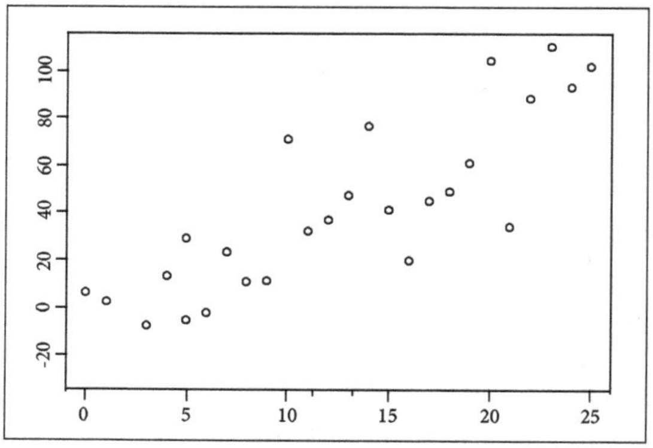

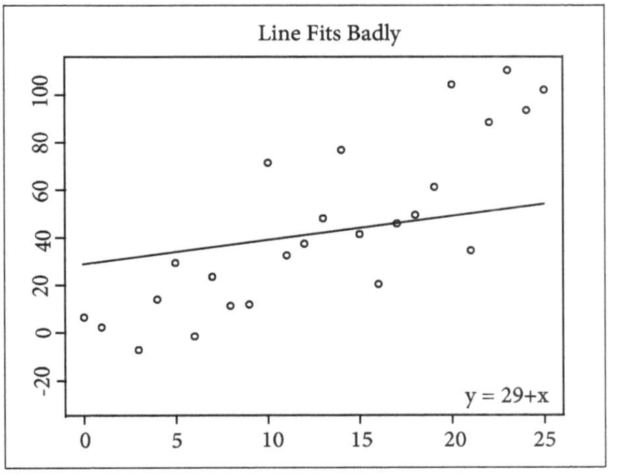

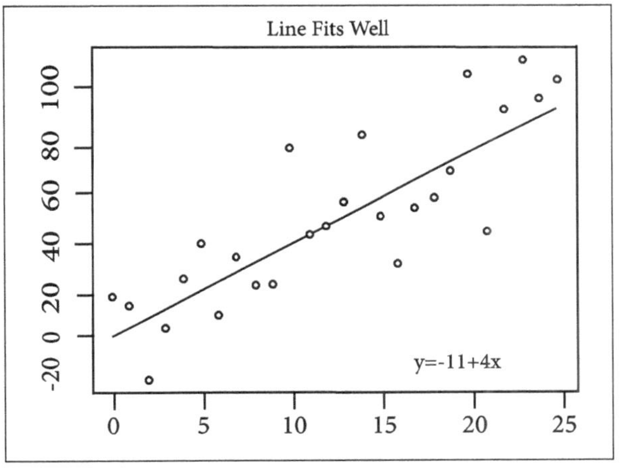

Fitting Criteria

Any assessment of how well a line fits a set of points must be based on how far the line deviates from the points,

$$d_i = y_i - (a + bx_i),\ i = 1,\dots.n$$

It makes sense to use the absolute deviations $|d_i|$ rather than the raw deviations d_i.

There are many different measures of how well a line fits a set of points.

Measures of Fit Quality

Sum of Absolute Deviations,

$$P(a,b) = \sum_{i=1}^{n} |y_i - (a + bx_i)|$$

Sum of Squared Deviations,

$$Q(a,b)=\sum_{i=1}^{n}\left|y_i-(a+bx_i)\right|^2$$

Maximum Deviation,

$$R(a,b) = \max\left|y_i - (a + bx_i)\right|$$

Least Squares

The most commonly used fitting criterion is that of least squares.

This means that we find the best fitting line by choosing a and b to minimize.

$$Q(a,b)=\sum_{i=1}^{n}\left|y_i-(a+bx_i)\right|^2$$

The justification for using this choice is that it produces the simplest statistical theory.

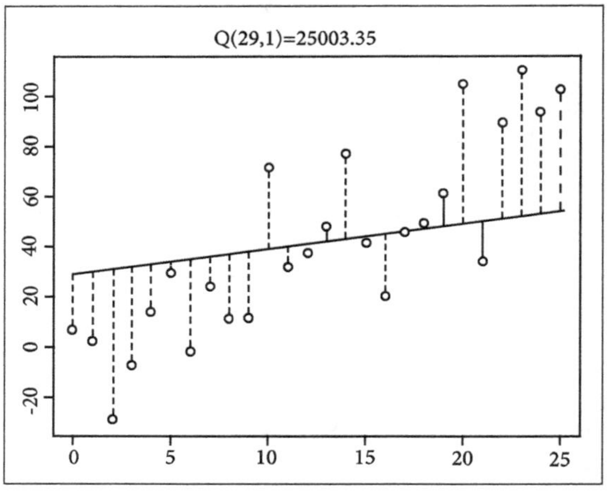

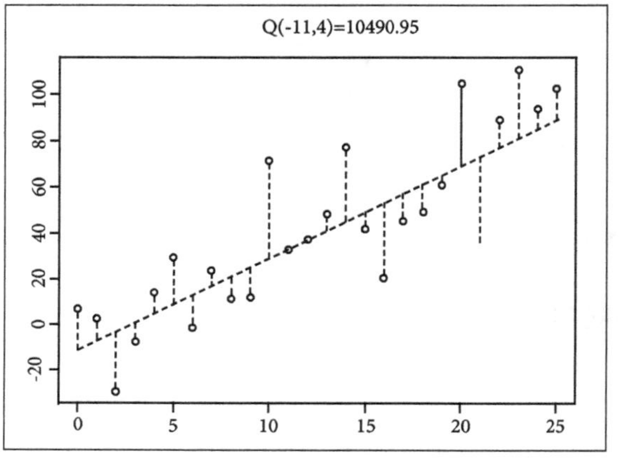

Finding the Best Slope and Intercept

There are a number of ways of finding the best fitting slope and intercept:

- The simplest method is exhaustive search.
- To carry out this method, we compute the value of Q(a,b) over a finely spaced grid.
- The results can be displayed with a contour plot.

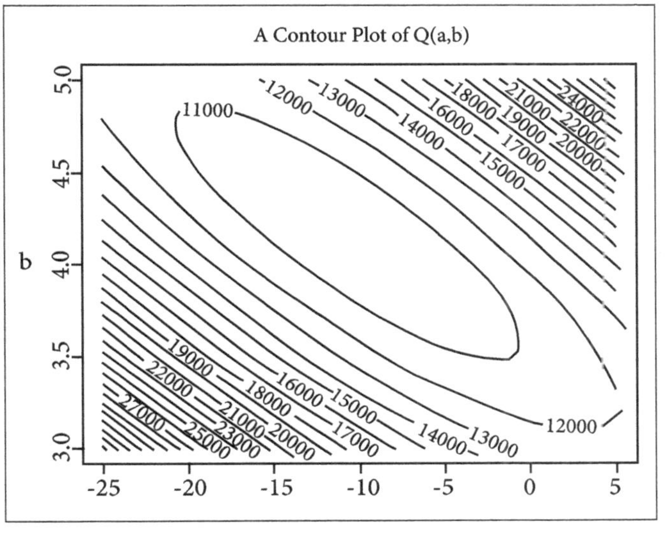

Precise Determination of a and b

The contour plots show that the best values of a and b are in the region of -10 and 4.2, but it is hard to be more precise:

- It is possible to finer and finer grids to zero-in on the best values, but it is possible to derive an exact formula for the best values.
- This will require a small diversion into mathematics.

A Simplified Problem

- Suppose we have data values $y_1,\ldots,y_n$ and we want to locate the point which minimizes,
- $Q(a)=\sum_{i=1}^{n}(y_i-a)^2$
- One way to proceed is to simply plot Q(a) as a function of a.
- In practise we compute Q(a) at a grid of points and we join up the dots.

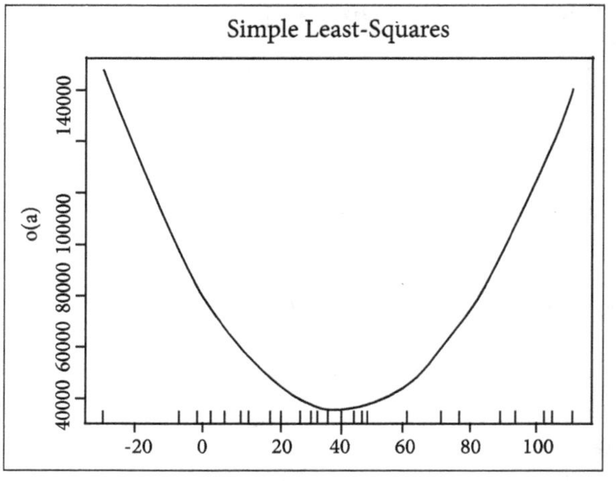

Formal Minimization

- Q(a) is a smooth function of a, so it can be minimized using calculus.
- We want the point where Q o (a) = o.

$$Q'(a) = \frac{d}{da}\sum_{i=1}^{n}(y_i - a)^2 = 2\sum_{i=1}^{n}(y_i - a)$$

The equation Q o (a) = o can be solved for a,

$$\sum_{i=1}^{n}(y_i - a) = 0,\quad \sum_{i=1}^{n} y_i = \sum_{i=1}^{n} a,\quad \sum_{i=1}^{n} y_i = na,$$

The Sample Mean

- We have just shown that y is the value of a which minimizes the function,

$$Q(a) = \sum_{i=1}^{n}(y_i - a)^2$$

- The sample mean is the solution of a least-squares minimization problem.

1.6.2 Correlation Analysis

In correlation analysis, we estimate a sample correlation coefficient, more specifically the Pearson Product Moment correlation coefficient. Sample correlation coefficient, denoted r, ranges between -1 and +1 and quantifies the direction and strength of the linear association between the two variables. The correlation between two variables can be positive (i.e., higher levels of one variable are associated with higher levels of the other) or negative (i.e., higher levels of one variable are associated with lower levels of the other).

The sign of the correlation coefficient indicates the direction of association. The magnitude of the correlation coefficient indicates strength of the association.

For example, a correlation of r = 0.9 suggests a strong, positive association between two variables, whereas a correlation of r = -0.2 suggest a weak, negative association. A correlation close to zero suggests no linear association between the two continuous variables.

It is important to note that there may be a non-linear association between two continuous variables, but computation of a correlation coefficient does not detect this. Therefore, it is always important to evaluate the data carefully before computing a correlation coefficient. The Graphical displays are particularly useful to explore associations between variables.

The figure below shows four hypothetical scenarios in which one continuous variable is plotted along the X-axis and the other along the Y-axis.

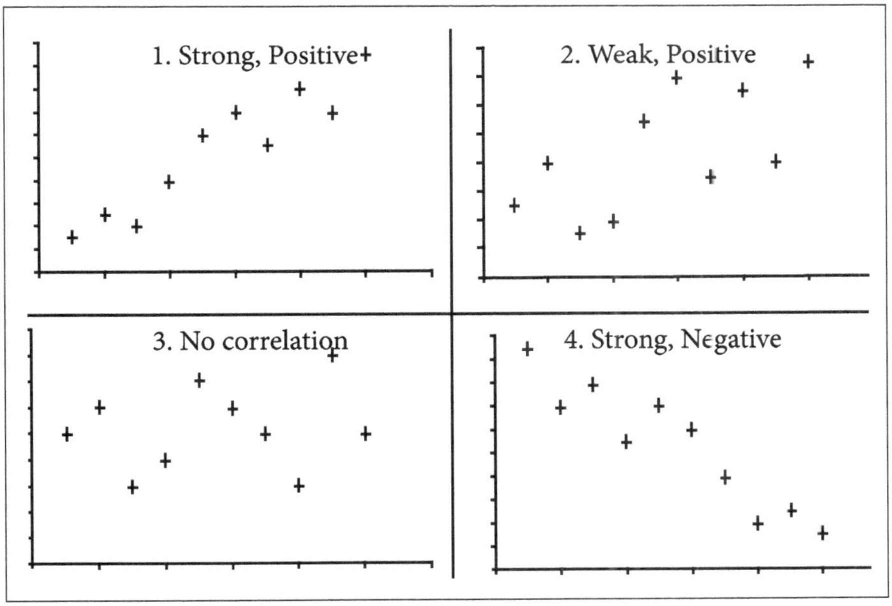

Scenario 1: It depicts a strong positive association (r=0.9), similar to what we might see for the correlation between infant birth weight and birth length.

Scenario 2: It depicts a weaker association (r = 0, 2) that we might expect to see between age and body mass index (which tends to increase with age).

Scenario 3: It might depict the lack of association (r approximately 0) between extent of media exposure in adolescence and age at which adolescents initiate sexual activity.

Scenario 4: It might depict the strong negative association (r = -0.9) generally observed between the number of hours of aerobic exercise per week and percent body fat.

Example: Correlation of Gestational Age and Birth Weight:

A small study is conducted involving 17 infants to investigate the association between gestational age at birth, measured in weeks and birth weight, measured in grams.

Infant ID #	Gestational Age (wks)	Birth Weight (gm)
1	34.7	1895
2	36.0	2030
3	29.3	1440
4	40.1	2835
5	35.7	3090
6	42.4	3827
7	40.3	3260
8	37.3	2690
9	40.9	3285
10	38.3	2920
11	38.5	3430
12	41.4	3657
13	39.7	3685
14	39.7	3345
15	41.1	3260
16	38.0	2680
17	38.7	2005

We wish to estimate the association between gestational age and infant birth weight. In this example, birth weight is the dependent variable and gestational age is the independent variable. Thus y=birth weight and x=gestational age. The data are displayed in a scatter diagram in the figure.

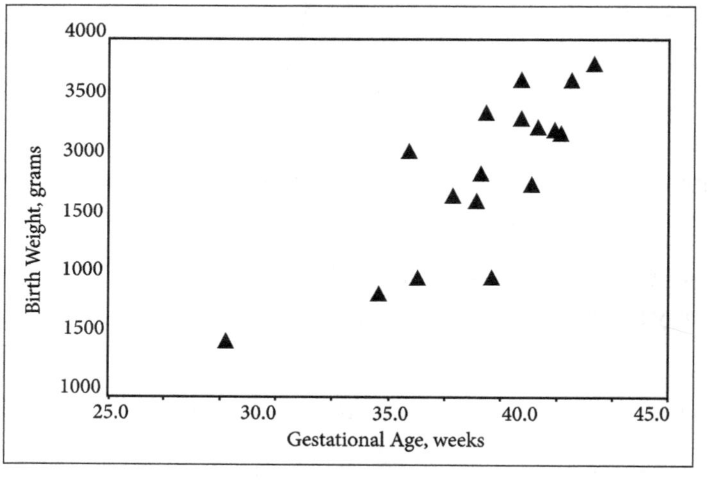

Each point represents an (x, y) pair (in this case the gestational age, measured in weeks and the birth weight, measured in grams). Note that the independent variable is on the horizontal axis (or X-axis) and the dependent variable is on the vertical axis (or Y-axis). The scatter plot shows a positive or direct association between gestational age and birth weight. Infants with shorter gestational ages are more likely to be born with lower weights and infants with longer gestational ages are more likely to be born with higher weights.

The formula for the sample correlation coefficient is,

$$r = \frac{\mathrm{Cov}(x,y)}{\sqrt{S_x^2 * S_y^2}}$$

Where Cov(x, y) is the covariance of x and y defined as,

$$\mathrm{Cov}(x,y) = \frac{\sum(X-\bar{X})(Y-\bar{Y})}{n-1}$$

S_x^2 and S_y^2 are the sample variances of x and y, defined as,

$$S_x^2 = \frac{\sum(X-\bar{X})^2}{n-1} \quad \text{and} \quad S_y^2 = \frac{\sum(Y-\bar{Y})^2}{n-1}$$

The variances of x and y measure the variability of the x scores and y scores around their respective sample means ($\bar{X}$ and $\bar{Y}$, considered separately). Covariance measures the variability of the (x, y) pairs around the mean of x and mean of y, considered simultaneously.

To compute the sample correlation coefficient, we need to compute the variance of gestational age, the variance of birth weight and also covariance of gestational age and birth weight.

We first summarize the gestational age data. The mean gestational age is,

$$\bar{X} = \frac{\sum X}{n} = \frac{652.1}{17} = 38.4.$$

To compute the variance of gestational age, we need to sum the squared deviations (or differences) between each observed gestational age and the mean gestational age. The computations are summarized below:

Infant ID #	Gestational Age	$(X-\bar{X})$	$(X-\bar{X})^2$
1	34.7	-3.7	13.69

2	36.0	-2.4	5.76
3	29.3	-9.1	82.81
4	40.1	1.7	2.89
5	35.7	-2.7	7.29
6	42.4	4.0	16.00
7	40.3	1.9	3.61
8	37.3	-1.1	1.21
9	40.9	2.5	6.25
10	38.3	-0.1	0.01
11	38.5	0.1	0.01
12	41.4	3.0	9.00
13	39.7	1.3	1.69
14	39.7	1.3	1.69
15	41.1	2.7	7.29
16	38.0	-0.4	0.16
17	38.7	0.3	0.09
	$\sum X = 652.1$	$\sum(X-\bar{X}) = 0$	$\sum(X-\bar{X}) = 159.45$

The variance of gestational age is,

$$S_x^2 = \frac{\sum(X-\bar{X})^2}{n-1} = \frac{159.45}{16} = 10.0.$$

Next, we summarize the birth weight data. The mean birth weight is,

$$\bar{Y} = \frac{\sum Y}{n} = \frac{49{,}334}{17} = 2902.$$

The variance of birth weight is computed just as we did for gestational age as shown in the table below:

Infant ID #	Birth Weight	$(Y-\bar{Y})$	$(Y-\bar{Y})^2$
1	1895	-1007	1.014.049
2	2030	-872	760,384
3	1440	-1462	2.137.444
4	2835	-67	4,489
5	3090	188	35.344
6	3827	925	855,625
7	3260	358	128.164

8	2690	-212	44.944
9	3285	383	146.689
10	2920	18	324
11	3430	528	278.784
12	3657	755	570,025
13	3685	783	613.089
14	3345	443	196,249
15	3260	358	128.164
16	2680	-222	49,284
17	2005	-897	804.609
	$\Sigma Y = 49{,}334$	$\sum(Y-\bar{Y})=0$	$\sum(Y-\bar{Y})^2 = 7{,}767{,}660$

The variance of birth weight is,

$$S_y^2 = \frac{\sum(Y-\bar{Y})^2}{N-1} = \frac{7{,}767{,}660}{16} = 485{,}578.8.$$

Next we compute the covariance,

$$\text{Cov}(x,y) = \frac{\sum(X-\bar{X})(Y-\bar{Y})}{n-1}$$

To compute the covariance of gestational age and birth weight, we need to multiply the deviation from mean gestational age by the deviation from the mean birth weight for each participant i.e.,

$$(X-\bar{X})(Y-\bar{Y})$$

The computations are summarized below. Notice that we simply copy the deviations from the mean gestational age and birth weight from the two tables above into the table below and multiply:

Infant Identification Number	$(X-\bar{X})$	$(Y-\bar{Y})$	$(X-\bar{X}XY-\bar{Y})$
1	-3.7	-1007	3725.9
2	-2.4	-872	2092.8
3	-9.1	-1462	13,304.2
4	1.7	-67	-113.9
5	-2.7	188	-507.6
6	4.0	925	3700.0

7	1.9	358	680.2
8	-1.1	-212	233.2
9	2.5	383	957.5
10	-0.1	18	-1.8
11	0.1	528	52.8
12	3.0	755	2265.0
13	1.3	783	1017.9
14	1.3	443	575.9
15	2.7	358	966.6
16	-0.4	-222	88.8
17	0.3	-897	-269.1
			$\sum(X-\bar{X})(Y-\bar{Y})=28{,}468.4$

The covariance of gestational age and birth weight is,

$$S_y^2 = \frac{\sum(Y-\bar{Y})^2}{n-1} = \frac{7{,}767{,}660}{16} = 485{,}578.8.$$

We now compute the sample correlation coefficient,

$$r = \frac{Cov(x,y)}{\sqrt{S_x^2 * S_y^2}} = \frac{1798.0}{\sqrt{10.0 * 485{,}578.8}} = \frac{1798.0}{2199.4} = 0.82$$

Not surprisingly, the sample correlation coefficient indicates a strong positive correlation.

As we noted, sample correlation coefficients range from −1 to +1. In practice, meaningful correlations (i.e., correlations that are clinically or practically important) can be as small as 0.4 (or -0.4) for positive (or negative) associations. There are also statistical tests to determine whether an observed correlation is statistically significant or not (i.e., statistically significantly different from zero).

2

Partial Differential Equation

2.1 Partial Differential Equation of First Order

The general form of a first order partial differential equation is,

$$F(x, y, z, p, q) = 0 \quad ...(1)$$

Where,

x, y = Two independent variables

z = Dependent variable

p = zx

q = zy

A solution or integral of the p.d.e is the relation between the dependent and independent variable satisfying the equation.

Complete Solution

Any function $f(x, y, z, a, b) = 0 \quad ...(2)$

Involving two arbitrary constants a, b and satisfying p.d.e (1) is known as the complete solution or complete integral or primitive. Geometrically the complete solution represents a two parameter family of surfaces.

Example: $(x + a)(y + b) = z$ is the complete solution of the p.d.e $z = pq$.

Particular Solution

A solution obtained by giving particular value to the arbitrary constants in the complete solution is called a particular solution of the p.d.e. It represents a particular surface of a family of surfaces given by the complete solution.

Example: $(x + 3)(y + 4) = z$ is the particular solution of the p.d.e $z = pq$.

General Solution

In the complete solution if we put b = φ (a) then we get a solution containing an arbitrary function φ, which is called a general solution. It represents the envelope of the family of surfaces,

$$f(x,y,z\,\phi\,(a)) = 0.$$

Singular Solution

Differentiating the p.d.e (1) w.r.t the arbitrary constants a and b we get,

$$\frac{\partial F}{\partial a} = 0 \text{ and } \frac{\partial F}{\partial b} = 0$$

Suppose it is possible to eliminate a and b from the three equations then the relation so obtained is called the singular solution of the p.d.e. Singular solution represents the envelope of the two parameter family of surfaces.

Example: The complete solution of the p.d.e z = pq is (x + a) (y + b)

Differentiating partially w.r.t a and b we get

$$x + a = 0 \text{ and } y + b = 0. \Rightarrow z = 0 \text{ is the singular solution.}$$

Problems

1. Solve the equation $\frac{\partial^2 z}{\partial x \partial y} = x^2 y.$

Solution:

The given equation can be written as $\frac{\partial}{\partial x}\left(\frac{\partial z}{\partial y}\right) = x^2 y$

Integrating w.r.t. x we get,

$$\frac{\partial}{\partial y} = \frac{x^3 y}{3} + f(y)$$

Where f(y) is an arbitrary function of y,

Integrating the above w.r.t. x we get,

$$z = \frac{x^3 y^2}{6} + \int f(y)dy + g(x)$$

$$z = \frac{1}{6}(x^3 y^2) + F(y) + g(x)$$

Where g (x) is an arbitrary function of x.

This is the general solution of the given equation.

2. Solve the equation $xy\frac{\partial^2 z}{\partial x \partial y} - x\frac{\partial z}{\partial x} = y^2$.

Solution:

Diving throughout by x the equation may be written as,

$$y\frac{\partial}{\partial y}\left(\frac{\partial z}{\partial x}\right) - \frac{\partial z}{\partial x} = \frac{y^2}{x}$$

$$y\frac{\partial p}{\partial y} - p = \frac{y^2}{X} \text{ or } \frac{1}{y}\frac{\partial p}{\partial y} - \frac{p}{y^2} = \frac{1}{x}$$

$$\Rightarrow \frac{\partial}{\partial y}\left(\frac{p}{y}\right) = \frac{1}{x}$$

Integrating the above w.r.t. y we get,

$$\frac{p}{y} = \frac{1}{x}y + f(x)$$

$$\frac{1}{y}\frac{\partial z}{\partial x} = \frac{y}{x} + f(x)$$

Integrating the above w.r.t. x we get,

$$\frac{z}{y} = y\log x + \int f(x)dx + g(y)$$

$$\Rightarrow z = y^2 \log x + yF(x) + G(y)$$

3. Solve: $\frac{\partial^2 z}{\partial y^2} - x\frac{\partial z}{\partial y} = -\sin y - x\cos y$.

Solution:

We know that $q = \frac{\partial z}{\partial y}$, the given equation may be written as,

$$\frac{\partial q}{\partial y} - xq = -\sin y - x\cos y \qquad ...(1)$$

Since x is treated as constant, this equation is a first order ordinary linear d.e in which q is the dependent variable and y is the independent variable. For this equation,

$$I.F = e^{\int -x dy} = e^{-xy}$$

∴ The solution of (1) is

$$qe^{-xy} = \int(-\sin y - x\cos y)e^{-xy}dy + f(x)$$

$$= \int \frac{d}{dy}(\cos y)e^{-xy}dy + f(x)$$

$$qe^{-xy} = e^{-xy}\cos y + f(x)$$

or,

$$\frac{\partial z}{\partial y} = \cos y + e^{xy}f(x)$$

Integrating w.r.t y, we get,

$$z = y + \frac{e^{xy}}{x}f(x) + g(y)$$

2.1.1 Linear Partial Differential Equation

The general first order linear partial differential equation has the form,

$$a(x,t)\frac{\partial v(x,t)}{\partial x} + b(x,t)\frac{\partial v(x,t)}{\partial t} + c(x,t)v(x,t) + d(x,t) \qquad ...(1)$$

Where, a, b, c and d are the given functions. These functions are assumed to be continuously differentiable. At each point (x, t) where the vector [a(x, t), b(x, t)] is defined and non-zero, the left hand side of equation (1) represents a directional derivative of v(x, t) in the direction of [a(x, t), b(x, t)].

Equation (2) determines a family of curves x = x(s), t = t(s).

$$\frac{dx}{ds} = a(x,t), \quad \frac{dt}{ds} = b(x,t) \qquad ...(2)$$

The tangent vector to the curve x = x(s), t = t(s) is [x'(s), t'(s)]. Equation (2) states that the tangent vector coincides with the direction of the vector [a(x, t), b(x, t)] at each point, where [a(x, t), b(x, t)] is defined and non-zero.

Therefore, the derivative of v(x, t) along these curves become,

$$\frac{dv}{ds} = \frac{dv[x(s),t(s)]}{ds} = \frac{\partial v}{\partial x}\frac{dx}{ds} + \frac{\partial v}{\partial t}\frac{dt}{ds} = a\frac{\partial v}{\partial x} + b\frac{\partial v}{\partial t} = cv + d \quad ...(3)$$

The family of curves x = x(s), t = t(s), v = v(s) determined by solving the system of ODES of equation (2) and (3), are called as the characteristic curves of the PDE. Since equation (2) can be solved independently of equation (3), the curves in the (x, t) plane determined from equation (1) are sometimes referred to as the characteristic curves or characteristic base curves.

The approach we develop to solve equation (1) by using the solutions of equation (2) and (3) is called as the method of characteristics. As shown, it is based on the geometric interpretation of the partial differential equation.

The existence and uniqueness theory for ordinary differential equations, assuming certain smoothness conditions on the functions a, b, c and d, guarantees that exactly one solution curve (x(s), t(s), v(s)) of equation (2) and (3) passes through a given point (x_0, t_0, v_0) in (x, t, v) space. As a rule, we are determining a specific solution, v = v(x, t) that passes through or contains a given curve C. This problem is known as the initial value problem for equation (1).

The initial value problem for equation (1) is solved by the method of characteristics as follows. We represent the initial curve C parametrically as,

$$x = x(\tau), \qquad t = t(\tau), \qquad v = v(\tau) \qquad ...(4)$$

For a given range of values of the parameter T.

The curve may be of finite or infinite extent and must have a continuous tangent vector at each point. Every value of τ fixes a point on C through which a unique characteristic curve passes. The family of characteristic curves determined by the points of C may be parameterized as,

$$x = x(s, \tau), \qquad t = t(s, \tau), \qquad v = v(s, \tau) \qquad ...(5)$$

Where, s = 0 corresponds to the initial curve C, (i.e.) we have x(0, τ) = x(τ), t(0, τ) = t(τ) and v(0, τ) = v(τ). In general, equation (5) yields a parametric representation of a surface in (x, t, v) space that contains the initial curve C. Assuming that the equations x = x(s, τ), t = t(s, τ) can be inverted to give s and τ as functions of x and t, these functions can be introduced into the equation v = v(s, τ).

The resulting function v = V (x, t) satisfies equation (1) in a neighborhood of the curve C in view of equation (3), the initial condition of equation (4) and is the unique solution of the given initial value problem. It is referred to as the integral surface for the problem. The smoothness requirements placed on the functions a, b, c and d in equation (3) imply that V (x, t) is continuously differentiable near the curve C.

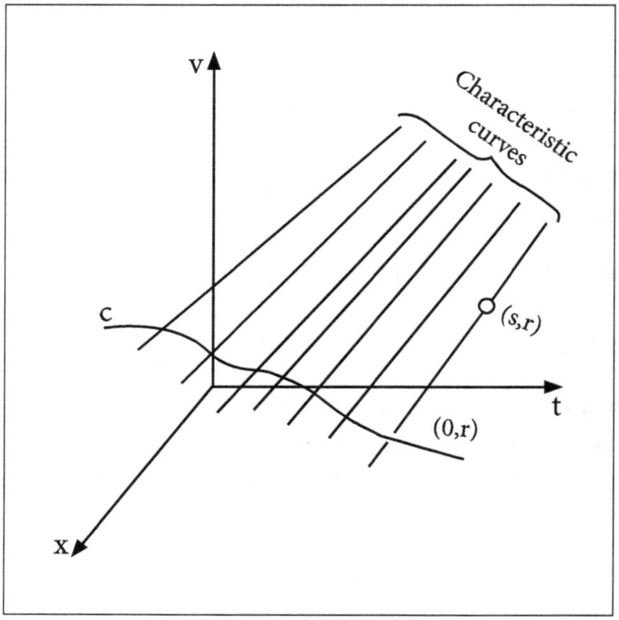

Construction of the integral surface.

The above figure shows how the integral surface is constructed in terms of the initial and characteristic curves. If the foregoing method does not result in an integral surface, the initial value problem may not have a solution at all or it may have infinitely many solutions. The latter situation arises if the initial curve C is itself a characteristic curve. Even if an integral surface can be constructed, it may be discontinuous or fail to be differentiable along some curve.

2.1.2 Non-linear Partial Differential Equation

The general non-linear first order PDE is written in the form,

$$F(x, y, z, z_x, z_y) = 0 \qquad ...(1)$$

Where, F is not linear in zx and zy. Setting zx = p and zy = q, equation (1) can be rewritten as,

$$F(x, y, z, p, q) = 0 \qquad ...(2)$$

The method of characteristics for non-linear PDEs:

$$F(x, y, z, p, q) = a(x, y)p + b(x, y)q + c(x, y)z - d(x, y) = 0$$

In this method, PDE becomes an ODE along the characteristic curve, which may be regarded as the solution of the system,

$$x'(t) = a(x(t), y(t)) \text{ and } y'(t) = b(x(t), y(t)) \quad ...(3)$$

Note that, Fp = a(x, y) and Fq = b(x, y). Hence, (3) may be written as,

$$x'(t) = F_p \text{ and } y'(t) = F_q \quad ...(4)$$

For solving the first order non-linear PDE of equation (1), equation (4) defines the characteristic curve as the solution of the system,

$$x'(t) = F_p(x(t), y(t), z(t), p(t), q(t)) \text{ and } y'(t)$$

$$= F_q(x(t), y(t), z(t), p(t), q(t)) \quad ...(5)$$

Where, z(t) = z(x(t), y(t)), p(t) = z_x(x(t), y(t)), q(t) = z_y(x(t), y(t)). However, unlike the linear case, the right hand sides of equation (5) depend not only on x(t) and y(t), but also on z(t), p(t) and q(t). Thus, we can expect a large system of five ODEs for the five unknown x(t), y(t), z(t), p(t) and q(t).

For the remaining three equations, note that,

$$z'(t) = \frac{d}{dt}\{z(x(t), y(t))\}$$

$$z'(t) = z_x x'(t) + z_y y'(t)$$

$$z'(t) = p(t)x'(t) + q(t)y'(t)$$

$$z'(t) = p(t)F_p(x(t), y(t), z(t), p(t), q(t))$$

$$+ q(t)F_q(x(t), y(t), z(t), p(t), q(t)) \quad ...(6)$$

2.2 Homogeneous and Non-Homogeneous Partial Differential Equation with Constant Co-Efficient

Second order non-homogeneous linear differential equation with constant co-efficient is of the form,

$$ay'' + by' + cy = G(x) \quad ...(1)$$

Where, a, b and are constants and is a continuous function. The related homogeneous equation,

$$ay'' + by' + cy = 0 \qquad ...(2)$$

Is known as the complementary equation and plays an important role in the solution of the original non-homogeneous equation (1).

Theorem:

The general solution of the non-homogeneous differential equation (1) can be written as,

$$y(x) = y_p(x) + y_c(x)$$

Where y_p is a particular solution of equation (1) and y_c is the general solution of the complementary equation (2).

Solution of homogeneous Partial differential equations involving derivative with respect to one independent variable only:

In the process of solution, we make use of the method of solving ordinary linear differential equations.

Problems

1. Let us solve the equation $\frac{\partial^2 z}{\partial x^2} - a^2 z = 0$ under the condition $z = 0$ and $\frac{\partial z}{\partial x} = a \sin y$ when $x = 0$.

Solution:

Given:

$$\frac{\partial^2 z}{\partial x^2} - a^2 z = 0$$

Using the D-Operator, we write the above equation as,

$$\left(D^2 - a^2\right)z = 0 \text{ where } D = \frac{\partial}{\partial x}$$

Here we treat y as constant. Then this equation is an ordinary second order linear homogeneous d.e. in which x is the independent variable and z is the dependent variable.

$\therefore$ The A.E is $m^2 - a^2 = 0 \rightarrow m = \pm a$

$\therefore$ G.S is $z = Ae^{ax} + Be^{-ax}$

Where A and B are arbitrary function of y,

Now,

$$\frac{\partial Z}{\partial X} = Aae^{ax} - Bae^{-ax}$$

Using the given condition, i.e., z =0 and $\frac{\partial Z}{\partial X} = a \sin y$ when x =0.

$$a \sin y = Aa - Ba = a(A - B)$$

$$0 = A + B$$

$$A - B = \sin y$$

$$A + B = 0$$

Solving the above simultaneous equations, we get,

$$2A = \sin y$$

$$A = (1/2) \sin y$$

$$B = -(1/2) \sin y$$

The solution is,

$$z = \frac{1}{2}\left(e^{ax} - e^{-ax}\right)\sin y$$

$$z = \sin h \, ax \sin y$$

2. Let us solve the equation $\frac{\partial^2 z}{\partial x^2} - 2\frac{\partial z}{\partial x} + 2z = 0$ Given that $z = e^y$ and $\frac{\partial z}{\partial x} = 0$ when $x = 0$.

Solution:

Given:

The given equation can be put in the form,

$$\left(D^2 - 2D + 2\right)z = 0 \text{ where } D = \frac{\partial}{\partial x}$$

If y is treated as constant, then the above equation is a O.D.E with the A and B,

$$\left(m^2 - 2m + 2\right) = 0$$

$$m = 1 \pm i$$

$\therefore$ The G.S is $z = e^x (A \cos x + B \sin x)$

Where A and B are arbitrary functions of y,

$$\frac{\partial z}{\partial x} = e^x (A\cos x + B\sin x) + e^x (-A\sin x + B\cos x)$$

$$\frac{\partial z}{\partial x} = e^x \left[(A+B)\cos x + (B-A)\sin x\right]$$

But,

$$\frac{\partial z}{\partial x} = 0$$

$$0 = B + A$$

And,

$z = e^y$, when $x = 0$.

$$e^y = A$$

$$B = -e^y$$

$$\therefore\ z = e^x \left[e^y \cos x - e^y \sin x\right]$$

$$z = e^{x+y} [\cos x - \sin x]$$

3. Let us solve the equation $\frac{\partial^2 u}{\partial x \partial y} = \frac{\partial u}{\partial x}$ using the substitution $\frac{\partial u}{\partial x} = v$.

Solution:

Given:

We know that,

$$\frac{\partial^2 u}{\partial x \partial y} \quad \frac{\partial^2 u}{\partial y \partial x}$$

Given PDE can be written as,

$$\frac{\partial v}{\partial y} = v, \quad \text{where} \quad v = \frac{\partial u}{\partial x}$$

or,

$$\frac{\partial v}{\partial y} - v = 0$$

By considering it as ODE we can write it as,

$$\frac{dv}{dy} - v = 0 \qquad \rightarrow (D-1)\,\upsilon = 0$$

A.E is $m - 1 = 0 \rightarrow m = 1$

$\therefore$ The solution is given by $U = f(x)\, e^{y}$

i.e., $\frac{\partial u}{\partial x} = f(x)e^{y}$

On integrating with respect to x,

we get,

$$u = F(x)\, e^{y} + g(y)$$

Where,

$$F(x) = \int f(x).dx$$

2.2.1 Cauchy Type and Monge's Method

Consider homogeneous linear differential equation as,

$$a^{n}\frac{d^{n}y}{dx^{n}} + a_{1}x^{n-1}\frac{d^{n-1}y}{dx^{x-1}} + \ldots.. + a_{n}y = X \ldots\ldots\ldots\ldots(1)$$

(Here a's are constants and x be a function of X) is known as Cauchy's homogeneous linear equation is the homogeneous linear equation with variable coefficients,

$$a_{n}x^{n}\frac{d^{n}y}{dx^{n}} + a^{n-1}\frac{d^{n-1}}{dx^{n-1}} + \ldots. + a_{1}x_{0}\frac{dy}{dx} + ay = f(x)$$

It is also known as Euler's equation.

Equation (1) can be transformed into a linear equation of constant Coefficients by the transformation.

$$x = e^{z}$$

or,

$$z = \log x$$

Then,

$$Dy = \frac{dy}{dx} = \frac{dy}{dz}\frac{dz}{dx} = \frac{1}{x}\frac{dy}{dz}$$

$$\left[D = \frac{d}{dx}, \theta = \frac{d}{dz}\right]$$

$$x\frac{d}{dx} = xD = \frac{d}{dz} = \theta$$

Similarly,

$$\frac{d^2y}{dx^2} = \frac{d}{dx}\left(\frac{dy}{dx}\right) = \frac{d}{dx}\left(\frac{1}{x}\frac{dy}{dz}\right)$$

$$= -\frac{1}{x^2}\frac{dy}{dz} + \frac{1}{x}\frac{d}{dx}\left(\frac{dy}{dz}\right) = -\frac{1}{x^2}\frac{dy}{dz} + \frac{1}{x}\frac{d}{dz}\left(\frac{dy}{dz}\right)\frac{dz}{dx}$$

$$\frac{d^2y}{dx^2} = -\frac{1}{x^2}\frac{dy}{dz} + \frac{1}{x^2}\frac{d^2y}{dz^2}$$

$$\therefore x^2\frac{d^2y}{dx^2} = \frac{d^2y}{dz^2} - \frac{dy}{dz}$$

$$\therefore x^2D^2 = \left(\theta^2 - \theta\right) = \theta\left(\theta - 1\right)$$

Similarly,

$$x^3D^3 = \theta\left(\theta - 1\right)\left(\theta - 2\right)$$

$$x^4D^4 = \theta\left(\theta - 2\right)\left(\theta - 3\right)$$

Similarly,

$$\therefore xD = \theta$$

$$x^2D^2 = \theta\left(\theta - 1\right)$$

$$x^3D^3 = \theta\left(\theta - 1\right)\left(\theta - 2\right)$$

and so on.

Monge-Ampere Equation

Moose's method gives the solution of partial differential equation,

$$Rr + Ss + Tt = V \qquad ...(1)$$

Where R, S, T and V are functions of x, y, z, p, q.

This method reduces the equation (1) into an equivalent system of two equations from which we determine p or q or both p and q. If p or q is determined then solution is obtained following the procedure of solving Lagrange equation and if p and q both are determined then solution is obtained by integrating dz = pdx + qdy.

We have,

$$dp = \frac{\partial p}{\partial x}dx + \frac{\partial p}{\partial y}dy = rdx + sdy$$

$$dq = \frac{\partial q}{\partial x}dx + \frac{\partial q}{\partial y}dy = sdx + tdy$$

From these equations,

$$r = \frac{dp - sdy}{dx}, \quad t = \frac{dq - sdx}{dy}$$

Substitute these values in (1),

$$R\frac{dp - sdy}{dx} + Ss + T\frac{dq - sdx}{dy} = V$$

$$(Rdpdy + Tdqdx - Vdxdy) - s(R(dy)^2 - Sdydx + T(dx)^2) = 0$$

This equation holds for arbitrary values of s and hence,

$$R(dy)^2 - sdydx + T(dx)^2 = 0$$

$$Rdpdy + Tdqdx - Vdxdy = 0 \qquad ...(2)$$

And,

Equations (1) and (2) are called Mange's auxiliary equations or Monge's subsidiary equations.

From equation, (2)

$$dy = \frac{1}{2R}\left[S \pm \sqrt{s^2 - 4RT}\right]dx$$

Now, dy has two non-zero values when $S^2 - 4RT \neq 0, R \neq 0, T \neq 0$

Using these two values of dy and equation (3), we try to find two relations between x, y, z, p, q, each relation containing an arbitrary function. These relations are called intermediate integrals. We solve these relations for p and q and then we integrate.

$$dz = pdx\ qdy.$$

To obtain the solution of given partial differential equation. We may also use only one of the values of dy from two values obtained and then use the method of solving the first order partial differential equations like Lagrange's equation to obtain the solution. For this purpose, we may use dz = pdx qdy.

In equation (2),

$$S^2 - 4RT = 0\ (\text{or})\ R = 0\ (\text{or})\ T = 0$$

Then dy will have only one non-zero value.

In this case, we use the method of solving the first order partial differential equation like Lagrange's equation to obtain the solution. We may use dz = pdx + qdy.

Problems

1. Let us solve $x^2\frac{d^2y}{dx^2} + x\frac{dy}{dx} + y = 4\sin\left(\log x\right)$.

Solution:

Given:

$$x^2\frac{d^2y}{dx^2} + x\frac{dy}{dx} + y = 4\sin\left(\log x\right)$$

Consider the transformation,

$$x = e^z\left(\text{or}\right)\ z = \log x$$

$$\therefore\ xD = \theta$$

$$x^2D^2 = \theta\left(\theta - 1\right)$$

$$\left(x^2D^2 + xD + 1\right)y = 4\ \sin\left(\log x\right)$$

$$\left(\theta^2 - 1\right)y = 4\sin(z)$$

R.H.S = 0

$$\left(\theta^2 - 1\right)y = 0$$

A.E,

$m^2 + 1 = 0, m = \pm i$

C.F = A cosz + B sinz

C.F = A cos(log x) + B sin (log x)

$$P.I = \frac{1}{\theta^2 + 1} 4\,(\sin\ z)$$

$$= 4\left(-\frac{z\cos\ z}{2}\right) = -2z\cos z$$

P.I = -2 log x cos(log x)

Complete solution is,

y = A cos (log x) + B sin (log x) - 2logx cos(log x)

y = (A - 2log x) cos (log x) - 2logx cos(log x).

2. Here we shall solve $x^2\frac{d^2y}{dx^2} + 4x\frac{dy}{dx} + 2y = x\log x.$

Solution:

Given:

$$x^2\frac{d^2y}{dx^2} + 4x\frac{dy}{dx} + 2y = x\log x$$

$$\left(x^2D^2 + 4xD + 2\right)y = x\log x \qquad ...(1)$$

Consider,

$$x = e^z\,(or)\ z = \log x$$

$$\therefore\ xD = \theta$$

$$x^2D^2 = \theta(\theta - 1)$$

Equation (1) becomes,

$$(\theta(\theta-1)+4\theta\ +2)y = e^z z$$

$$(\theta^2+3\theta\ +\ 2)y = ze^z$$

$$A.E : m^2+3m+2=0$$

$$M=-2,\ -1$$

$$C.F = Ae^{-2z}+Be^{-z} = Ae^{\log x^{-2}}+Be^{\log x^{-1}}$$

$$C.F = \frac{A}{x^2}+\frac{B}{x}$$

$$P.I = \frac{1}{(\theta^2+3\theta+2)}(e^z z)$$

$$= e^z\frac{1}{(\theta+1)^2+3(\theta+1)+2}z$$

$$= e^z\frac{1}{(\theta^2+5\theta+6)}z$$

$$= \frac{e^z}{6}\left(1+\left(\frac{\theta^2+5\theta}{6}\right)\right)^{-1} z$$

$$= \frac{e^z}{6}\left(1-\frac{5}{6}\theta\right)z = \frac{e^z}{6}\left(z-\frac{5}{6}\right)$$

$$= \frac{e^{\log x}}{6}\left(\log x-\frac{5}{6}\right)$$

$$= \frac{x}{6}\left(\log x-\frac{5}{6}\right)$$

Complete solution is,

$$y = C.F + P.I$$

$$= \frac{A}{x^2}+\frac{B}{x}+\frac{x}{6}\left(\log x-\frac{5}{6}\right)$$

2.3 Second Order Partial Differential Equation

Problems

1. Let us solve the partial differential equation $(D^2 - 2DD' + D'^2)\, z = 0$.

Solution:

Auxiliary equation is $m^2 - 2m + 1 = 0,\ m = 1,1.$

$\therefore$ Solution is $Z = f(y+x) + x\, g(y+x)$.

Where,

$$D = \frac{\partial}{\partial x} \text{ and } D' = \frac{\partial}{\partial y}.$$

2. Let us solve the linear partial differential equation of second order $(D^2 + 2DD' + D'^2 - 2D - 2D')z = \sin(x + 2y)$, where $D = \frac{\partial}{\partial x}$ and $D' = \frac{\partial}{\partial y}$.

Solution:

$$\left(D^2 + 2DD' + D'^2 - 2D - 2D'\right)z = \sin(x+2y)$$

$$\text{(i.e.)}\ (D+D')(D+D'-2)z = \sin\ (x+2y) \qquad \ldots(1)$$

Solution of (1) is $z = C.F + P.I$

$$C.F = \phi_1(y-x) + e^{2x}\phi_2(y-x)$$

$$P.I = \frac{1}{\left(D^2 + 2DD' + D'^2 - 2D - 2D'\right)}\sin(x+2y)$$

$$= \frac{1}{(-1-4-2D-2D')}\sin(x+2y)$$

Put,

$$D^2 = -1$$

$$= \frac{1}{-(2D+2D'+9)}\sin(x+2y) \qquad D'^2 = -4$$

$$= \frac{1}{-(2D+2D'+9)}\sin(x+2y) D'^2 = -4$$

$$= \frac{-[(2D+2D')-9]}{-[(2D+2D')+9][(2D+2D')-9]} \sin(x+2y)$$

$$P.I = \frac{(2D+2D'-9)\sin(x+2y)}{-\left[(2D+2D')^2-81\right]}$$

$$= \frac{(2D+2D'-9)\sin(x+2y)}{-\left[4D^2+8DD'+4D'^2-81\right]}$$

$$= \frac{2\cos(x+2y)+4\cos(x+2y)-9\sin(x+2y)}{-117}$$

$$P.I = \frac{}{39}\left[2\cos(x+2y)-3\sin(x+2y)\right]$$

$$z = \phi_1(y-x)+e^{2x}\phi_2(y-x) + \frac{1}{39}\left[2\cos(x+2y)-3\sin(x+2y)\right].$$

3. Let us solve the differential equation of higher order (D^2 – 3DD' + 2D'^2 + 2D – 2D)'z = x + y + sin (2x + y).

Solution:

$$\left(D^2-3DD'+2D'^2+2D-2D\right)'z = x+y+\sin(2x+y).$$

$$(D-D')(D-2D'+2)z = (x+y)\sin(2x+y) \qquad ...(1)$$

Solution of (1) is z = C. F + P.I.

Now,

$$C.F = f_1(y+x)+e^{-2x}f_2(y+2x)$$

$$P.I = P.I_1 + P.I_2$$

$$P.I_1 = \frac{1}{(D-D')(D-2D'+2)}(x+y)$$

$$= \frac{1}{2D}\left(1-\frac{D'}{D}\right)^{-1}\left(1+\frac{D-2D'}{2}\right)^{-1}(x+y)$$

$$= \frac{1}{2D}\left(1-\frac{D'}{D}\right)^{-1}\left[1-\left(\frac{D-2D'}{2}\right)+\left(\frac{D-2D'}{2}\right)\ldots\right](x+y)$$

$$= \frac{1}{2D}\left[1+\frac{D'}{D}+\frac{D'^2}{D^2}+\ldots\right]\left[(x+y)-\frac{1}{2}+1\right]$$

$$= \frac{1}{2D}\left[\left(x+y+\frac{1}{2}\right)+\frac{1}{D}\right]$$

$$= \frac{1}{2D}\left[x+y+\frac{1}{2}+x\right]$$

$$P.I_1 = \frac{1}{2D}\left[2x+y+\frac{1}{2}\right] = \frac{1}{2}\left[x^2+xy+\frac{x}{2}\right]$$

$$P.I_2 = \frac{\sin(2x+y)}{D^2-3DD'+2D'^2+2D-2D'}$$

$$= \frac{\sin(2x+y)}{-4+6-2+2(D-D')}$$

$$= \frac{\sin(2x+y)}{2(D-D')} = \frac{(D+D')\sin(2x+y)}{2(D^2-D'^2)}$$

$$= -\frac{1}{6}\left[2\cos(2x+y)+\cos(2x+y)\right]$$

$$= -\frac{1}{2}\cos(2x+y)$$

The general solution is,

$$Z = f_1(y+x)+e^{-2x} f_2(y+2x)$$

$$+\frac{1}{2}\left(x^2+xy+\frac{x}{2}\right)-\frac{1}{2}\left[\cos(2x+y)\right].$$

4. Let us solve the partial equation $(D^2 - D'^2)z = e^{x-y}\sin(2x + 3y)$.

Solution:

$$\left(D^2 - D^2\right)z = e^{x-y}\sin(2x+3y) \quad ...(1)$$

Solution of (1) is Z = C.F + P.I

$$\text{A.E } m^2 - 1 = 0, (\text{i.e.})\, m = 1 \text{ twice}$$

$$\text{C.F} = (f_1 + x) + x f_2 (y + x)$$

$$= \frac{e^{x-y}\sin(2x+3y)}{\left(D^2 - D'^2\right)}$$

$$= e^{x-y} \frac{\sin(2x+3y)}{\left[(D+1)^2 - (D'-1)^2\right]}$$

P.I

Replace $D^2 = -4$

$D' = -9$

P.I

$$= e^{x-y} \frac{\sin(2x+3y)}{-4+2D+9+2D'}$$

$$= e^{x-y} \frac{\sin(2x+3y)}{\left[(2D+2D')+5\right]}$$

$$= e^{x-y} \frac{\left[(2D+2D')-5\right]\sin(2x+3y)}{\left[(2D+2D')+5\right]\left[(2D+2D')-5\right]}$$

$$= e^{x-y} \frac{\left[(2D+2D')-5\right]\sin(2x+3y)}{(2D+2D')^2 - 25}$$

$$= e^{x-y} \frac{\left[(2D+2D')-5\right]\sin(2x+3y)}{4D^2 + 4D'^2 + 8DD' - 25}$$

$$P.I = e^{x-y}\frac{[(2D+2D')-5]\sin(2x+3y)}{-16-36-48-25}$$

$$= -\frac{e^{x-y}}{125}[4\cos(2x+3y)+6\cos(2x+3y)-5\sin(2x+3y)]$$

$$= -\frac{e^{x-y}}{125}[10\cos(2x+3y)-5\sin(2x+3y)]$$

$$= \frac{e^{x-y}}{25}[\sin(2x+3y)-2\cos(2x+3y)]$$

General solution is,

$$Z = f_1(y+x)+xf_2(y+x)+$$

$$\frac{e^{x-y}}{25}[\sin(2x+3y)-2\cos(2x+3y)].$$

5. Let us solve the differential equation $(2D^2 - 5DD' + 2D'^2) z = 5 \sin (2x + y)$.

Solution:

The A.E is $2m^2 - 5m + 2 = 0$

$$m = \tfrac{1}{2}, 2$$

The C.F is $= \phi_1\left(y+\frac{x}{2}\right)+\phi_2(y+2x)$

$$PI = \frac{1}{2D^2-5DD'+2}5\sin(2x+y)$$

$$= -\frac{5x}{6}D\sin(2x+y)$$

$$= \frac{5x}{6}\times 2\cos(2x+y)$$

$$= -\frac{5x}{3}\cos(2x+y)$$

6. Let us solve the partial equation of second order (D² - D) (z) = e^{x-y} sin(x + 2y).

Solution:

$$\left(D^2 - D\right)(z) = e^{x-y}\sin(x+2y) \qquad ...(1)$$

Solution of (1) is,

$$z = C.F + P.I \qquad ...(2)$$

A.E.

$$m^2 - 1 = 0$$

Solving m = ± 1

$$C.F. = f_1(y+x) + f_2(y-x) \qquad ...(3)$$

$$P.I = \frac{e^{x-y}\sin(x+2y)}{D^2 - D'^2}$$

$$= e^{x-y}\frac{\sin(x+2y)}{(D+1)^2 - (D'-1)^2}$$

$$= e^{x-y}\frac{\sin(x+2y)}{D^2 - D'^2 + 2D + 2D'},$$

Put,

$$D^2 = -1$$

$$= e^{x-y}\frac{\sin(x+2y)}{(2D+2D'+3)} D'^2 = -4$$

$$P.I = e^{x-y}\frac{\sin(x+2y)}{\left[2(D+D')+3\right]}$$

$$= e^{x-y}\frac{\left[2(D+D')-3\right]\sin(x+2y)}{\left[2(D+D')+3\right]\left[2(D+D')-3\right]}$$

$$= e^{x-y}\frac{\left[2D+2D'-3\right]\sin(x+2y)}{4\left(D^2 + D'^2 + 2DD'\right) - 9}$$

$$= e^{x-y}\frac{\left[2\cos(x+2y)+4\cos(x+2y-3\sin(x+2y))\right]}{4(-9)-9}$$

$$= \frac{e^{x-y}}{-45}(6\cos(x+2y)-3\sin(x+2y))$$

$$= \frac{e^{x-y}}{15}(\sin(x+2y)-2\cos(x+2y))$$

Solution of (1) is,

$$Z = f_1(y+x)+f_2(y-x)+\frac{e^{x-y}}{15}$$

$$(\sin(x+2y)-2\cos(x+2y))$$

7. Let us solve the differential equation $(D^2 + 2DD' + D'^2)z = x^2y - e^{x-y}$.

Solution:

$$\left(D^2+2DD'+D'^2\right)z = x^2y+e^{x-y} \qquad ...(1)$$

Solution of (1) is,

$$Z = C.F + P.I \qquad ... (2)$$

AE,

$$m^2 + 2m + 1 = 0$$

$$(m+1)^2 = 0 \quad m = -1 \text{ twice}$$

$$C.F = f_1(y-X)+X\ f_2(y-x)$$

$$P.I = P.I_1 + P.I_2$$

$$P.I_1 = \frac{x^2y}{D^2+2DD'+D'^2}$$

$$= \frac{x^2y}{D^2\left[1+\left(\frac{2D'}{D}+\frac{D'^2}{D^2}\right)\right]}$$

$$=\frac{1}{D^2}\left[1+\left(\frac{2D'}{D}+\frac{D'^2}{D^2}\right)\right]^{-1}x^2y$$

$$=\frac{1}{D^2}\left[1-\left(\frac{2D'}{D}+\frac{D'^2}{D^2}\right)+\left(\frac{2D'}{D}+\frac{D'^2}{D^2}\right)^2-\ldots\right]x^2y$$

$$=\frac{1}{D^2}\left[1-\frac{2D}{D}\right]x^2y \qquad \because D'\,x^2y=x^2$$

$$=\frac{1}{D^2}\left[x^2y-\frac{2x^2}{D}\right] \qquad D^2x^2y=0$$

$$=\frac{1}{D^2}\left[x^2y-\frac{2x^3}{3}\right]$$

$$=\frac{1}{D}\left[\frac{x^3y}{3}-\frac{2x^4}{12}\right]$$

$$P.I_1=\frac{x^4y}{12}-\frac{x^5}{30}$$

$$P.I_2=\frac{e^{x-y}}{D^2+2DD'+D'^2}$$

Replace D = 1

$$=\frac{e^{x-y}}{1-2+1}D'=-1$$

$$=x\frac{e^{x-y}}{2D+2D'}$$

$$PI_2=\frac{x^2.e^{x-y}}{2}$$

$$Z=f_1(y-x)+xf_2(y-x)+\frac{x^4y}{12}-\frac{x^5}{30}+\frac{x^2e^{x-y}}{2}.$$

8. Let us solve the partial differential equation $(D^2 - 5DD' + 6D'^2)z = y \sin x$.

Solution:

Given:

$$\left(D^2 - 5DD' + 6D'^2\right)z = y \sin x \quad ...(1)$$

Solution of (1) is z = C.F + P.I ...(2)

AE,

$$m^2 - 5m + 6 = 0$$

$$(m-2)(m-3) = 0$$

$$m = 2, 3$$

$$C.F = f_1(y+2x) + f_2(y+3x) \quad ...(3)$$

$$P.I = \frac{y \sin x}{(D-2D')(D-3D')} \quad \left[P.I = \frac{1}{D-mD'} f(x)\right]$$

$$= \frac{1}{(D-2D')} \int [(C-3x)\sin dx\ dx = \int f(x),C-mx)dx]$$

Change,

$$y = C - 3x \text{ in } f(x,y)$$

$$P.I = \frac{1}{(D-2D')}\left[(C-3x)(-\cos x) - (-3)(-\sin x)\right]$$

$$= \frac{1}{(D-2D')}\left[(C-3x)(-\cos x) - 3 \sin x\right]$$

$$= \frac{-y \cos x}{(D-2D')} - \frac{3 \sin x}{(D-2D')}$$

$$= -\int (C-2x)\cos x\ dx - 3\int \sin x\ dx$$

$$= -\left[(C-2x)(\sin x) - (-2)(-\cos x)\right] + 3\cos x$$

$$= -\left[(y\sin x - \cos x) + 3\cos x\right]$$

$$P.I = 4\cos x - y\sin x.$$

$$Z = f1(y+2x) + f_2(y+3x) + 4\cos\sin x.$$

9. Let us solve the differential of equation $(D^3 - 7DD^2 - 6D'^3)z = \cos(x + 2y) + 4$.

Solution:

Solution of (1) is z =C.F + P.I ... (2)

Put D = m and D = 1 in (1) and equate to zero

$$m^3 - 7m - 6 = 0$$

Solving $(m + 1)(m^2 - m - 6) = 0$

i.e., $(m + 1)(m - 3)(m + 2) = 0$

The roots are $m_1 = -1$, $m_2 = 3$; $m_3 = -2$.

Hence the C.F is,

$$Z = \phi_1(y - X) + \phi_2(y+3x) + \phi_3(y-2x)(\text{use } \phi \text{ or } f)$$

$$P.I = P.I_1 + P.I_2$$

$$P.I_1 = \frac{\cos(x+2y)}{D^3 - 7DD'^2 - 6D'^3}$$

Replace $D^2 = -1$

$$= \frac{\cos(x+2y)}{-D + 28D + 24D'} \qquad D'^2 = -4$$

$$= \frac{\cos(x+2y)}{27D + 24D'} \qquad \text{and} \qquad DD' = -2$$

$$= \frac{D\cos(x+2y)}{27D^2 + 24DD'} \qquad D = \frac{\partial}{\partial x}$$

$$= \frac{D\cos(x+2y)}{-75} = \frac{-\sin(x+2y)}{-75}$$

$$P.I_1 = \frac{\sin(x+2y)}{75}$$

$$P.I_2 = \frac{4e^{ox+oy}}{D^3 - 7DD'^2 - 6D'^3}$$

Here Replace,

$$= 4x \frac{e^{ox+oy}}{3D^2 - 7D'^2} \qquad D = 0 \;\&\; D^1 = 0$$

Here,

$$Dr = 0$$

$$= 4x^2 \frac{e^{ox+oy}}{6D}$$

$$= 4x^3 \frac{e^{ox+oy}}{6}$$

$$P.I_2 = \frac{2x^3}{3}$$

∴Solution of (1) is,

$$z = \phi_1(y-x) + \phi_2(y+3x)$$

$$+\phi_3(y-2x) + \frac{\sin(x+2y)}{75} + \frac{2x^3}{3}.$$

10. Let us solve the partial equation of higher order $(D^2 - D'^2 - 3D + 3D')z = xy + 7$.

Solution:

$$D^2 - D^2 - 3D + 3D' = (D+D')(D-D') - 3(D-D')$$
$$= (D-D')(D+D'-3)$$

∴ The given equation becomes,

$$(D-D')(D+D'-3)Z - xy + 7 \qquad ...(1)$$

$$\left[\text{C.F for } (D - aD' - b) \text{ is } e^{bx} f(y+ax)\right]$$

$\therefore$ C.F $= f(y+x)e^{3x} f_2(y-x)$

$$\text{P.I} = \frac{(xy+7)}{(D-D')(D+D'-3)}$$

$$= \frac{(xy+7)}{(D-D')(-3)\left[1-\frac{(D+D')}{3}\right]}$$

$$= \frac{1}{(-3)(D-D')}\left[1-\left(\frac{D+D'}{3}\right)\right]^{-1}(xy+7)$$

$$= \frac{1}{(-3)(D-D')}\left[1+\left(\frac{D+D'}{3}\right)+\left(\frac{D+D'}{3}\right)^2+\ldots\right](xy+7)$$

$$= \frac{1}{-3D\left(1-\frac{D'}{D}\right)}$$

$$\left[1+\frac{D}{3}+\frac{D'}{3}+\frac{D^2}{9}+\frac{2DD'}{9}+\frac{D'^2}{9}+\frac{1}{27}\left(D^3+3D^2D'+3DD'^2+D'^3\right)+\ldots\right](xy+7)$$

$$= -\frac{1}{3}\left[\frac{1}{D}+\frac{D'}{D^2}\right]\left[1+\frac{D}{3}+\frac{D'}{3}+\frac{D^2}{9}+\frac{2DD'}{9}+\frac{D'^2}{9}+\frac{1}{7}\left(D^3+3D^2D'+3DD'^2+D'^3\right)+\ldots\right]$$

$$(xy+7)$$

$$= \frac{-1}{3}\left[\frac{1}{D}+\frac{1}{3}+\frac{D'}{3D}+\frac{D}{9}+\frac{2D'}{9}+\frac{DD'}{9}+\frac{D'}{D^2}+\frac{D'}{3D}+\frac{D'}{9}+\frac{DD'}{27}\right](xy+7)$$

$$= -\frac{1}{3}\left[\frac{x^2y}{2}+7x+\frac{xy}{3}+\frac{7}{3}+\frac{2x}{3D}+\frac{y}{9}+\frac{x}{3}+\frac{4}{27}+\frac{x}{D^2}\right]$$

$$\text{P.I} = -\frac{1}{3}\left[\frac{x^2y}{2}+7x+\frac{xy}{3}+\frac{67}{27}+\frac{x^2}{3}+\frac{y}{9}+\frac{x}{3}+\frac{x^3}{6}\right]$$

$\therefore$ Solution of (1) is,

$$Z = f(y+x)+e^{3x} f_2(y-x)-\frac{1}{3}$$

$$\left[\frac{x^2}{2}+7x+\frac{xy}{3}+\frac{67}{27}+\frac{x^2}{3}+\frac{y}{9}+\frac{x}{3}+\frac{x^3}{6}\right].$$

11. Let us solve the partial equation $(D^2 + 2DD' + D'^2)z = \sin h\,(x + y) + e^{x+2y}$.

Solution:

Given:

$$\left(D^2+2DD'+D'^2\right)z=\sin\left(x+y\right)e^{x+2y}$$

$$=\frac{e^{(x+y)}}{2}-\frac{e^{-(x+y)}}{2}+e^{x+2y} \qquad ...(1)$$

$$\left(\because\ \sinh\ \theta=\frac{e^{\theta}-e^{-\theta}}{2}\right)$$

Solution of (1) is,

$$Z=C.F+P.I \qquad ...(2)$$

$$m^2+2m+1=0$$

Auxiliary equation,

Solving,

$$\left(m+1\right)=0\Rightarrow m=-1 \text{ twice}$$

$$\therefore\ C.F=f_1\left(y-x\right)+xf_2\left(y-x\right)$$

$$P.I=P.I_1+P.I_2+P.I_3$$

$$P.I_1=\frac{1}{2}\frac{e^{x+y}}{\left(D^2+2DD'+D'^2\right)}$$

Put,

$$D=1$$

And,

$$D'=1$$

$$P.I_1 = \frac{e^{x+y}}{8}$$

$$P.I_2 = -\frac{1}{2}\frac{e^{-x-y}}{D^2 + 2DD' + D'^2}$$

Put,

$$D = -1; \ D' = -1$$

$$= -\frac{e^{-(x+y)}}{8}$$

$$P.I_3 = \frac{e^{x+2y}}{D^2 + 2DD' + D'^2}$$

Put,

$$D = 1\Omega \ D' = 2$$

$$= \frac{e^{x+2y}}{9}$$

$$\therefore \text{Sol.}: Z = f_1(y-x) + xf_2(y-x) + \frac{1}{8}\left(e^{x+y} - e^{-(x+y)}\right) + \frac{e^{x+2y}}{9}.$$

12. Let us solve the partial differential equation [2D² - DD' - D '² + 6D + 3D'] z = x eʸ.

Given:

$$\left(2D^2 - DD' - D^2 + 6D + 3D'\right)Z = x\,e^y \qquad ...(1)$$

Solution:

Now,

$$2D^2 - DD' - D'^2 + 6D + 3D'$$

$$= (2D + D')(D - D') + 3(2D + D')$$

$$= (2D + D')(D = D' + 3)$$

∴Equation (1) can be written as,

$$(2D+D')(D-D'+3)z=xe^{y} \qquad ...(2)$$

$$C.F = Z = f_1\left(y-\frac{x}{2}\right)+e^{-3x} f_2(y+x)+\frac{e^{y}}{2}\left(x-\frac{5}{2}\right).$$

P.I

$$=f_1\left(y-\frac{x}{2}\right)+e^{-3x} f_2(y+x)$$

$$=\frac{1}{2D^2-DD'-D'^2+6D+3D'}xe^{y}$$

$$=e^{y}\frac{1}{2D^2-D(D'+1)-(D'+1)^2+6D+3(D'+1)}x$$

$$=e^{y}\frac{1}{2D^2-DD'-D-D'^2-2D'-1+6D+3D'+3}x$$

$$=\frac{e^{y}}{2}\frac{1}{\left[1+\frac{1}{2}\left(5D+D'+2D^2-DD'-D'^2\right)\right]}x$$

$$=\frac{e^{y}}{2}\left[1+\frac{1}{2}\left(5D+D'+2D^2-DD'-D'^2\right)\right]^{-1}x$$

$$=\frac{e^{y}}{2}\left[1-\frac{5}{2}D\right]x \qquad \therefore D\ x=0;\ D'x=0$$

$$=\frac{e^{y}}{2}\left[x-\frac{5}{2}\right]$$

∴ Solution of (1) is,

$$Z=f_1(y-x)+f_2(y-2x)+f_3(y+2x)-\frac{x}{12}\cos(2x+y).$$

2.4 The Vibrating String

Let us consider small transverse vibrations of an elastic string of length l. which is stretched and then fixed at its two ends. Now we will study the transverse vibration of the string when no external forces act on it. Take an end of the string as the origin and the string in the equilibrium position as the x-axis and the line through the origin and perpendicular to the x-axis as the y-axis. We make the following assumptions:

- The motion takes place entirely in one plant. This plane is chosen as the xy plane.
- In this plane. each panicle of the string moves in a direction perpendicular to the equilibrium position of the string.
- The tension T caused by stretching the string before fixing it at the end points is constant at all times at all points of the deflected string.
- The tension T is very large compared with the weight of the string and hence the gravitational force may be neglected.
- The effect of friction is negligible.
- The string is perfectly flexible. It can transmit only tension but not bending or shearing forces.
- The slope of the deflection curve is small at all points and at all times.

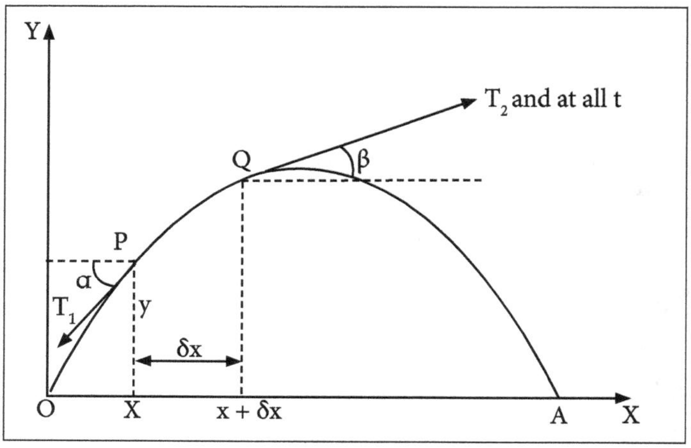

When the string is in motion in the xy-plan. the displacement y of any point of the string is a function of x and time t. Let P (x. y) and Q ($x + \delta x$, $y + \delta y$) be two neighbouring points on the string. Let a and β be the inclinations made by the tangents at P and Q respectively with the x-axis. The tension T_1 at P and tension T_2 at Q arc tangential.

Let m be the mass per unit length of the string which is homogeneous. Consider the motion of the infinitesimal element PQ of the string.The vertical component of the force to which this element is subjected to is given equation.

$$T_1 \cos\alpha = T_2 \cos\beta = T(\text{constant}) \qquad ...(1)$$

Dividing (1) by T, we get,

$$\frac{m\,\delta s \dfrac{\partial^2 y}{\partial t^2}}{T} = \frac{T_2 \sin\beta}{T} - \frac{T_1 \sin\alpha}{T}$$

$$\frac{m\delta s}{T}\frac{\partial^2 y}{\partial t^2} = \frac{T_2 \sin\beta}{T_2 \cos\beta} - \frac{T_1 \sin\alpha}{T_1 \cos\alpha} \qquad \begin{bmatrix} T = T_1 \cos\alpha \\ T = T_2 \cos\beta \end{bmatrix}$$

$$\Rightarrow \frac{m\delta s}{T}\frac{\partial^2 y}{\partial t^2} = \tan\beta - \tan\alpha \Rightarrow \frac{\partial^2 y}{\partial t^2} = \frac{T}{m\delta s}(\tan\beta - \tan\alpha)$$

Slope $\tan\alpha = \left(\frac{\partial y}{\partial x}\right)_x$ at P and slope $\tan\beta = \left(\frac{\partial y}{\partial x}\right)_{x+\delta x}$ at Q.

$$\frac{\partial^2 y}{\partial t^2} = \frac{T}{m\delta x}\left[\left(\frac{\partial y}{\partial x}\right)_{x+\delta x} - \left(\frac{\partial y}{\partial x}\right)_x\right] \Rightarrow \frac{\partial^2 y}{\partial t^2} = \frac{T}{m}\left[\frac{\left(\frac{\partial y}{\partial x}\right)_{x+\delta x} - \left(\frac{\partial y}{\partial x}\right)_x}{\delta x}\right]$$

On taking limit as $\delta x \to 0$,

We have,

$$\frac{\partial^2 y}{\partial t^2} = \frac{T}{m}\frac{\partial^2 y}{\partial x^2}, \qquad \boxed{\frac{\partial^2 y}{\partial t^2} = a^2 \frac{\partial^2 y}{\partial x^2}} \qquad \left(\text{Take } a^2 = \frac{T}{m}\right)$$

2.4.1 The Wave Equation and its Solution

Let us derive the boundary conditions and initial conditions for solving the vibration of string equation if the string is subjected to initial displacement f(x) and initial velocity g(x).

The wave equation is expressed as,

$$\frac{\partial^2 y}{\partial t^2} = a^2 \frac{\partial^2 y}{a x^2}$$

The boundary conditions are as follows:

$$\left.\begin{array}{l}(1)\ y\,(o,t)=o\\(2)\ y\,(l,t)=o\end{array}\right\}t\geq o$$

$$\left.\begin{array}{l}(3)\ \dfrac{\partial y}{\partial t}(x,o)=g(x)\\(4)\ \ y(x,o)=f(x)\end{array}\right\}o<x<1$$

Problems

1. A string is stretched and fastened to two point's l apart. Motion is started by displacing the string into the form y = k (lx - x²) from which it is released at time t = o. Let us determine the displacement of any point of the string at a distance x from one end at any time t.

Solution:

Given:

$$y=k\left(lx-x^2\right)$$

The equation is,

$$\frac{\partial^2 y}{\partial r^2}=a^2\frac{\partial^2 y}{\partial x^2}$$

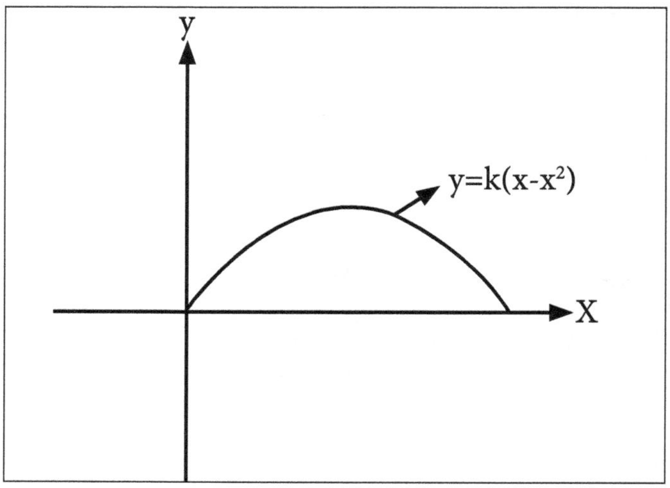

Here the Boundary conditions are,

- $y(o,t)=o;\ t\geq o$

- $y(l,\ t)=0;\ t\geq 0$
- $\frac{\partial y}{\partial t}l_{t=0}=0;\ 0\leq x\leq l$
- $y(x,\ 0)=k\left(1x-x^{2}\right);0\leq x\leq l$

The suitable solution of the one dimensional wave equation is,

$$y(x,t)=(A\cos\lambda x+B\sin\lambda x)(C\cos\lambda\, at+D\sin\lambda\, at)$$

After using conditions (i), (ii) and (iii) to the above solution we get,

$$y(x,t)=B\sin\left(\frac{n\pi x}{1}\right)$$

The general solution is,

$$y(x,t)=\sum_{n=1}^{\infty}B_{n}\sin\left(\frac{n\pi x}{1}\right)\cos\left(\frac{n\pi ar}{1}\right)$$

Where,

$$B_{n}=BC.$$

Using the condition (iv) to the above equation, we have,

$$y(x,0)=\sum_{n=1}^{\infty}B_{n}\sin\left(\frac{n\pi x}{1}\right),0\leq x\leq 1.$$

$$\therefore B_{n}=\frac{2}{l}\int_{0}^{1}K\left(lx-x^{2}\right)\sin\left(\frac{n\pi x}{1}\right)dx$$

$$=\frac{4kl^{2}}{n^{3}\pi^{3}}\left[1-(-1)^{n}\right]$$

∴ The solution is given by,

$$y(x,t)=\sum_{n=1}^{\infty}\frac{4kl}{n^{3}\pi^{3}}\left[1-(-1)^{n}\right]\sin\left(\frac{n\pi x}{1}\right)\cos\left(\frac{n\pi ar}{1}\right).$$

2. A tightly stretched string of length 'l' has its ends fastened at x = 0 and x = 1. The mid-point of the string is then taken to height 't' and released from rest in that position. Let us find the lateral displacement of a point of the string at time 't' from the instant of release.

Solution:

Given:

A tightly stretched string of length 'l' has its ends fastened at x = 0 and x = 1. The mid-point of the string is then taken to height 't' and released from rest in that position.

The one dimensional wave equation is given by,

$$\frac{\partial^2 y}{\partial t^2} = a^2 \frac{\partial^2 y}{\partial x^2}.$$

The possible solutions are,

$$y(x,t) = \left(A_1 e^{px} + A_2 e^{-px}\right)\left(A_3 e^{pat} + A_4 e^{-pat}\right)$$

$$y(x,t) = \left(A_5 \cos px + A_6 \sin px\right)\left(A_7 \cos pat + A_8 \sin pat\right)$$

$$y(x,t) = \left(A_9 x + A_{10}\right)\left(A_{11} t + A_{12}\right)$$

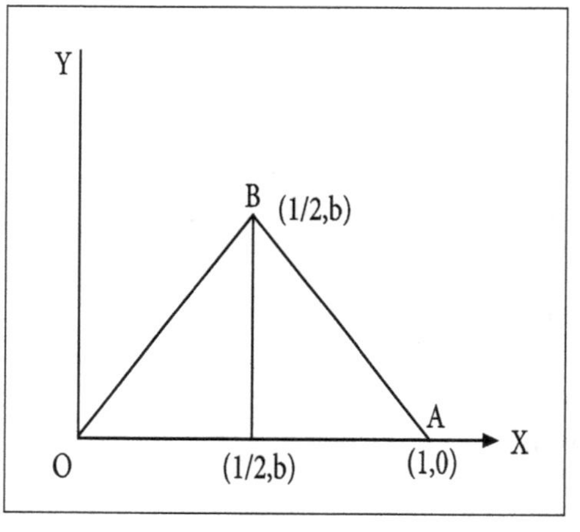

Initially the string is,

$$y(x,0) = \begin{cases} \text{equation of OB}; & 0 \le x \le \frac{1}{2} \\ \text{equation of BA}; & \frac{1}{2} \le x \le 1. \end{cases}$$

Boundary condition,

$$\left.\begin{aligned} y(0,t) &= 0 \\ y(l,t) &= 0 \end{aligned}\right\} t > 0t$$

$$\frac{\partial y}{\partial t}(x, o) = o$$

$$y(x,o) = \begin{cases} \frac{2b}{l}x; & o \le x \le \frac{l}{2} \\ \frac{2b}{l}(l-x); & \frac{l}{2} \le x \le l. \end{cases}$$

The suitable solution is,

$$y(x,t) = (A\cos px + B\sin px)(C\cos pat + D\sin pat) \qquad ...(1)$$

Applying (i) B.C in (1) put x = o

$$o = y(o,t) = A(C\cos pat + D\sin pat)$$

$$\Rightarrow A = o$$

Substituting in (1),

$$y(x,t) = B\ \sin\ px(C\ \cos\ pat + D\ \sin\ pat) \qquad ...(2)$$

Applying (ii) B.C in (2), put x = l

$$o = y(l,\ t) = B\sin pl\,(C\cos pat + D\sin pat)$$

$$\therefore\ \sin\ pl = o\,(\because \sin\theta = on\theta = n\pi)$$

$$\therefore\ pl = n\pi$$

$$\therefore p = \frac{n\pi}{l}$$

Substituting in (2),

$$y(x,t) = B\ \sin\frac{n\pi x}{l}\left(C\cos\frac{n\pi at}{l} + D\sin\frac{n\pi at}{l}\right) \qquad ...(3)$$

Differentiate both sides of (3) partially with respect to 't', we get,

$$\frac{\partial y}{\partial t}(x,t) = B\sin\frac{n\pi x}{l}\left[-\frac{n\pi a}{l}C\sin\frac{n\pi at}{l} + \frac{n\pi a}{l}D\cos\frac{n\pi at}{l}\right] \qquad ...(3a)$$

Applying (iii) B.C. in (3a), put t = o,

$$o = \frac{\partial y}{\partial t}(x, o) = \left(\frac{n\pi a}{l}\right)BD\ \sin\frac{n\pi x}{l}$$

$$\Rightarrow D = o$$

Substituting in (3),

$$y(x,t) = BC\ \sin\frac{n\pi x}{1}\cos\frac{n\pi at}{1} \qquad ...(4)$$

The most general solution,

$$y(x,t) = \sum_{n=1}^{\infty} B_n\ \sin\frac{n\pi x}{1}\cos\frac{n\pi at}{1}$$

Where,

$$BC = B_n$$

Applying B.C.(iv) in (I), put t = o

$$\left(f(x)\text{say}\right) = y(x,o) = \sum_{n=1}^{\infty} B_n\ \sin\frac{n\pi x}{1} \qquad ...(5)$$

Now we know that Half Range Fourier Sine series of f(x) is (o, 1),

$$f(x) = \sum_{n=1}^{\infty} b_n \sin\frac{n\pi x}{1} ...(6)$$

Where,

$$b_n = \frac{2}{l}\int_o^1 f(x)\sin\frac{n\pi x}{1}dx \qquad ...(7)$$

From equation (5) and (6), we get,

$$B_n = b_n = \frac{2}{l}\int_o^1 f(x)\sin\frac{n\pi x}{1}dx$$

$$B_n = \frac{2}{l}\left[\int_o^{1/2}\frac{2b}{l}x\sin\frac{n\pi x}{1}dx + \int_{1/2}^{1}\frac{2b}{l}(l-x)\sin\frac{n\pi x}{1}dx\right]$$

$$= \frac{4b}{l^2}\left[\left\{(x)\left(\frac{-1}{n\pi}\cos\frac{n\pi x}{1}\right) - (1)\left(\frac{-1^2}{n^2.\pi^2}\sin\frac{n\pi x}{1}\right)\right\}_o^{1/2} + \left\{(1-x)\left(\frac{-1}{n\pi}\cos\frac{n\pi x}{1} - (-1)\left(\frac{-l^2}{n^2\pi^2}\sin\frac{n\pi x}{1}\right)\right)\right\}_{1/2}^{1}\right]$$

$$=\frac{4b}{l^2}\left[\frac{-l^2}{n\pi}\cos\frac{n\pi}{2}+\frac{l^2}{n^2\pi^2}\sin\frac{n\pi}{2}+\frac{l^2}{n\pi}\cos\frac{n\pi}{2}+\frac{l^2}{n^2\pi^2}\sin\frac{n\pi}{2}\right]$$

$$=B_n=\frac{8b}{n^2\pi^2}\sin\frac{n\pi}{2}$$

Substituting in (I), we get,

$$y(x,t)=\frac{8b}{\pi^2}\sum_{n=1}^{\infty}\frac{1}{n^2}\sin\frac{n\pi}{2}\sin\frac{n\pi x}{l}\cos\frac{n\pi at}{l}$$

Is the required solution.

3. Let us consider a string of length c is initially at rest in its equilibrium position and each of its points is given a velocity v such that $v=\begin{cases}cx & \text{for } 0<x<\frac{l}{2}\\ c(l-x) & \text{for } \frac{l}{2}<x<l\end{cases}$. Also let us show that the displacement at any tie t is given by, $y(x,t)=\frac{4l^2c}{\pi^3a}\left[\sin\frac{\pi x}{l}\sin\frac{\pi at}{l}-\frac{1}{3^3}\frac{3\pi x}{l}\sin\frac{3\pi at}{l}+\ldots\right]$

Solution:

Given:

$$v=\begin{cases}cx & \text{for } 0<x<\frac{l}{2}\\ c(l-x) & \text{for } \frac{l}{2}<x<l\end{cases}$$

$$y(x,t)=\frac{4l^2c}{\pi^3a}\left[\sin\frac{\pi x}{l}\sin\frac{\pi at}{l}-\frac{1}{3^3}\frac{3\pi x}{l}\sin\frac{3\pi at}{l}+\ldots\right]$$

Let y(x, t) be the displacement then y(x, t) satisfies one dimensional wave equation,

$$\frac{\partial^2 y}{\partial x^2}=\alpha^2\frac{\partial^2 y}{\partial t^2}$$

With boundary conditions,

- $y(0, t) = 0$, for all $t \ge 0$
- $y(t, t) = 0$, for all $t \ge 0$
- $y(x, 0) = 0$, $0 \le x \le t$

- $$\frac{\partial y}{\partial t}(x,o)=\begin{cases} cx, & o \le x \le \frac{1}{2} \\ c(1-x), & \frac{1}{2} \le x \le 1 \end{cases}$$

The most suitable solution is,

$$y(x,t)=(A \cos px+B \sin px)(C \cos pat+D \sin pat) \qquad ...(1)$$

Substitute the first three boundary conditions in 1, we get,

$$y(x,t)=\sum_{n=1}^{\infty} b_n \sin\frac{n\pi}{l}x \sin\frac{n\pi}{l} at$$

Using condition (1), we get,

$$\frac{\partial y}{\partial t}(x,o)=\sum_{r=1}^{\infty} b_n \frac{n\pi}{l}\alpha \sin\frac{n\pi}{l}x$$

$$=\sum_{r=1}^{\infty} C_n \sin\frac{n\pi}{l}x$$

Where,

$$C_n=b_n\frac{n\pi a}{l}$$

$$C_n=\frac{2}{l}\int_o^t \frac{\partial y}{\partial t}(x,o)\sin\frac{n\pi}{l}xdx$$

Where,

$$b_n=\frac{4Cl^2}{n^3\pi^3 a}\sin\frac{n\pi}{2}$$

Hence the displacement is expressed as,

$$y(x,t)=\sum_{n=1}^{\infty}\frac{4l^2c}{\pi^3 a}\left[\sin\frac{n\pi}{2}\sin\frac{n\pi}{l}x\sin\frac{n\pi}{l}at\right]$$

4. A tightly stretched string of length 2I is fixed at both ends. The midpoint of the string is displaced by a distance “b” transversely and the string is released from rest in this position. Let us determine the displacement of any point of the string at any subsequent time.

Solution:

Given:

A tightly stretched string of length 2I is fixed at both ends. The midpoint of the string is displaced by a distance "b" transversely and the string is released from rest in this position.

The one dimensional wave equation is,

$$\frac{\partial^2 y}{\partial t^2} = a^2 \frac{\partial^2 y}{\partial x^2}$$

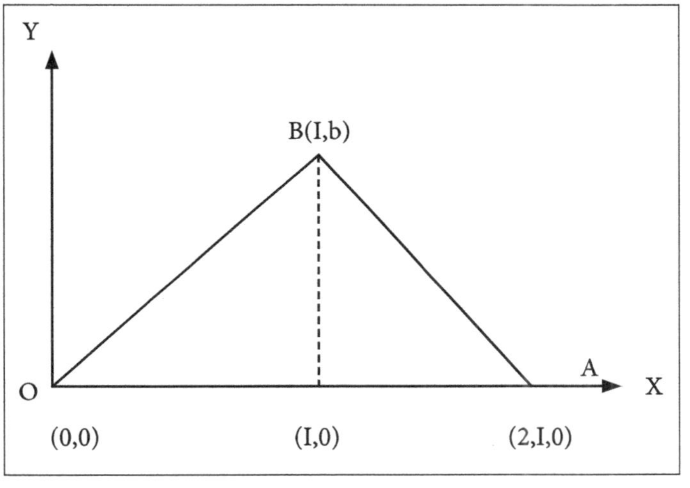

Initially, the string is given by,

$$y(x,o) = \begin{cases} \text{equation of OB}; \; o \le x \le l \\ \text{equation of BA}; \; l \le x \le 2l \end{cases}$$

$$= \begin{cases} \dfrac{b}{l}x; & o \le x \le l \\ \dfrac{b}{l} \times (2l - x); & o \le x \le 2l. \end{cases}$$

Consider 2l = L (for convenience). The Boundary conditions are,

$y(o, t) = o \; t > o$

$y(L, t) = o$

$$\frac{\partial y}{\partial t} = (x,o) = o$$

$$y(x,o)=\begin{cases}\frac{2b}{L}x;\ o\le x\le L/2\\ \frac{2b}{L}(L-x);\ L/2\le x\le L\end{cases}$$

The correct solution is,

$$y(x,t)=(A\cos px+B\sin px)(C\cos pat+D\sin pat) \qquad ...(1)$$

Using boundary condition (i) in (1), (Put x = o in (1))

$$o=y(o,t)=A(C\cos Pat+D\sin Pat)$$

$$\Rightarrow A=o$$

$$y(x,t)=B\sin Px\,(C\cos Pat+D\sin Pat) \qquad ...(2)$$

Applying (ii) B.C in (2)

Put,

$$x = L$$

$$o=y(L,t)=B\sin PL \qquad (C\cos Pat+P\sin Pat)$$

$$\Rightarrow \sin PL=o$$

$$\Rightarrow P=\frac{n\pi}{L}(\text{where } n=o,1,2,...,\infty$$

$$y(x,t)=B\sin\frac{n\pi x}{L}\left(C\cos\frac{n\pi at}{L}+D\sin\frac{n\pi at}{L}\right) \qquad ...(3)$$

Differentiate both sides of (3) partially with respect to we get,

$$\frac{\partial y}{\partial t}(x,t)=B\sin\frac{n\pi x}{L}\left(\frac{-n\pi a}{L}C\sin\frac{n\pi at}{L}+\frac{n\pi a}{L}D\cos\frac{n\pi at}{L}\right)$$

Applying B.C (iii),

$$o=\frac{\partial y}{\partial t}(x,o)=B\frac{(n\pi a)}{L}D\sin\frac{n\pi x}{L}$$

$$\Rightarrow D=o$$

$$\therefore y(x,t) = BC\sin\frac{n\pi x}{L}\cos\frac{n\pi at}{L} \qquad ...(4)$$

The most general solution of equation (1) is,

$$y(x,t) = \sum_{n=1}^{\infty} B_n \sin\frac{n\pi x}{L}\cos\frac{n\pi at}{L} \qquad ...(1)$$

Applying (iv) B.C Put t = o,

$$(\text{say } f(x)) = y(x,o) = \sum_{n=1}^{\infty} B\sin\frac{n\pi x}{L} \qquad ...(5)$$

We know that the Half range Fourier sine series of f(x) in (O,L),

$$f(x) = \sum_{n=1}^{\infty} b_n \sin\frac{n\pi x}{L} \qquad ...(6)$$

Where,

$$b_n = \frac{2}{L}\int_o^L f(x)\sin\frac{n\pi x}{L}dx \qquad ...(7)$$

From equation (5) and (6), we get,

$$B_n = b_n = \frac{2}{L}\int_o^L f(x)\sin\frac{n\pi x}{L}dx$$

$$= \frac{2}{L}\left[\int_o^{L/2}\frac{2b}{L}x\sin\frac{n\pi x}{L}dx + \int_{L/2}^{L}\frac{2b}{L}(L-x)\sin\frac{n\pi x}{L}dx\right]$$

$$= \frac{4b}{L^2}\left[\left\{(x)\left(\frac{-L}{n\pi}\cos\frac{n\pi x}{L}\right) - (1)\left\{\frac{-L^2}{n^2\pi^2}\sin\times\frac{n\pi x}{L}\right\}_o^{L/2}\right.\right.$$

$$\left.\left. + \left\{(L-x)\left(\frac{-L}{n\pi}\cos\frac{n\pi x}{L}\right) - (-1)\left(\frac{-L^2}{n^2\pi^2}\sin\frac{n\pi x}{L}\right)\right\}_{L/2}^{L}\right]$$

$$= \frac{4b}{L^2}\left[\frac{-L^2}{2h\pi}\cos\frac{n\pi}{2} + \frac{L^2}{n^2\pi^2}\sin\frac{n\pi}{2} + \frac{L^2}{2n\pi}\cos\frac{n\pi}{2} + \frac{L^2}{n^2\pi^2}\sin\frac{n\pi}{2}\right]$$

$$B_n = \frac{8b}{n^2\pi^2}\sin\frac{n\pi}{2}$$

Substituting in (1),

$$y(x,t)=\sum_{n=1}^{\infty}\frac{8b}{n^2\pi^2}\sin\frac{n\pi}{2}\sin\frac{n\pi x}{L}\cos\frac{n\pi at}{L}$$

Replace L by 2l, we get,

$$y(x,t)=\frac{8b}{\pi^2}\sum_{n=1}^{\infty}\frac{1}{n^2}\sin\frac{n\pi}{2}\sin\frac{n\pi x}{2l}\cos\frac{n\pi at}{2l}$$

Is the required solution.

6. A string of length l is initially at rest in its equilibrium position and each of its points is given a velocity v such that,

$$v=\begin{cases} cx; 0<x\le\frac{1}{2}. \\ c(1-x); \frac{1}{2}<x\le 1\end{cases}$$

Let us determine the displacement function y(x, t) at any time t.

Solution:

Given:

$$v=\begin{cases} cx; 0<x\le\frac{1}{2}. \\ c(1-x); \frac{1}{2}<x\le 1\end{cases}$$

The p.d.e giving vibrations of the string is,

$$\frac{\partial^2 y}{\partial t^2}=a^2\frac{\partial^2 y}{\partial x^2} \qquad ...(i)$$

The possibilities solution of equation (i) is,

$$y(x,t)=\left(C_1e^{px}+C_2e^{-px}\right)\left(C_3e^{pat}+C_4e^{-pat}\right)$$

$$y(x,t)=\left(C_5\cos px+C_6\sin px\right)\left(C_7\cos pat+C_8\sin pat\right)$$

$$y(x,t)=\left(C_9t+C_{10}\right)\left(C_{11}t+C_{12}\right)$$

The boundary conditions are,

$$y(0,t)=0;t\geq 0$$

$$y(l,t)=0;\ t\geq 0$$

$$y(x,0)=0;0\leq x\leq 1$$

$$\frac{\partial y}{\partial t}(x,0)=\begin{cases}cx;0<x\leq l/2\\ c(1-x);l/2<x\leq l.\end{cases}$$

The suitable solutions of (i) is of the form,

$$y(x,t)=(A\cos px+B\sin px)(C\cos\ pat+D\sin\ pat) \qquad ...(1)$$

Where A, B, C, D are arbitrary constants.

Putting x = 0 in(1),

$$0=y(0,t)=A(C\cos pat+D\sin pat)$$

$$\Rightarrow A=0$$

Substituting in (1), we get,

$$y(x,t)=B\sin px\,(C\cos pat+D\sin pat) \qquad ...(2)$$

Put x = l in (2) we get,

$$0=y(l,\ t)=B\ \sin\ pi\ (C\ \cos\ pat+D\ \sin\ pat)$$

$$\Rightarrow \sin\ pi=0$$

$$\Rightarrow pl=n\pi$$

$$\text{i.e.,}p=\frac{n\pi}{1}$$

Subs in (2), we get,

$$y(x,t)=B\sin\frac{n\pi x}{1}\left(C\cos\frac{n\pi at}{1}+D\sin\frac{n\pi at}{1}\right) \qquad ...(3)$$

Put,

t = 0 in (3) (Apply (ii) B.C)

$$o = y(x,o) = B\sin\frac{n\pi x}{1}(C)$$

$\Rightarrow C = o$ Substitute in (3)

$$y(x,f) = BD\sin\frac{n\pi x}{1}\sin\frac{n\pi at}{1}$$

Put,

$$BD = B_n$$

The most general solution will be of the form,

$$y(x,f) = \sum_{n=1}^{\infty} B_n \sin\frac{n\pi x}{1}\cdot\sin\frac{n\pi\, at}{1} \quad ...(A)$$

For applying in B.C, differentiate (A) partially with respect to 't',

$$\frac{\partial y}{\partial t}(x,t) = \sum_{n=1}^{\infty}\frac{n\pi a}{1} B_n \sin\frac{n\pi x}{1}\cos\frac{n\pi\, at}{1}$$

Put t = o,

$$f(x) = \frac{\partial y}{\partial t}(x,b) = \sum_{n=1}^{\infty}\frac{n\pi a}{1} B_n \sin\frac{n\pi x}{1} \quad ...(4)$$

We know that Half range Fourier sine series for f(x) in (o, l),

$$f(x) = \sum_{n=0}^{\infty} b_n \sin\frac{n\pi x}{1} \quad ...(5)$$

Where,

$$b_n \frac{2}{l}\int_0^1 f(x)\sin\frac{n\pi x}{1}dx \quad ...(6)$$

From equation (4) and (5), we get that,

$$\frac{n\pi a}{1}B_n = b_n = \frac{2}{l}\int_0^1 f(x)\sin\frac{n\pi x}{1}dx$$

$$= \frac{2}{l}\left[\int_0^{1/2} cx\sin\frac{n\pi x}{1}dx + \int_{1/2}^{1} c(1-x)\sin\frac{n\pi x}{1}dx\right]$$

$$\frac{n\pi a}{1} - B_n = \frac{2c}{l}\left[\left\{(x)\left(\frac{-1}{n\pi}\cos\frac{n\pi x}{1}\right) - (1)\left(\frac{-1^2}{n^2\pi^2}\cos\frac{n\pi x}{1}\right)\right\}_0^{1/2}\right.$$

$$\left. + \left\{(1-x)\left(\frac{-1}{n\pi}\cos\frac{n\pi x}{1}\right) - (-1)\left(\frac{-1^2}{n^2\pi^2}\sin\frac{n\pi x}{1}\right) - (-1)\left(\frac{-1^2}{n^2\pi^2}\sin\frac{n\pi x}{1}\right)\right\}_{1/2}^{1}\right]$$

$$= \frac{2c}{l}\left[\left\{\frac{-l^2}{2n\pi}\cos\frac{n\pi}{2} + \frac{1^2}{n^2\pi^2}\sin\frac{n\pi}{2}\right\} + \left\{(0) - \left(-\frac{l^2}{2n\pi}\cos\frac{n\pi}{2} - \frac{1^2}{n^2\pi^2}\sin\frac{n\pi}{2}\right\}\right]$$

$$= \frac{2c}{l}\left[\frac{2l^2}{n^2\pi^2}\sin\frac{n\pi}{2}\right]$$

$$\frac{n\pi a}{1} B_n = \frac{4cl}{n^2\pi^2}\sin\frac{n\pi}{2}$$

$$\Rightarrow B_n = \frac{4cl^2}{n^2\pi^3 a}\sin\frac{n\pi}{2} \text{ Substitude in } (A)$$

$$y(x,t) = \sum_{n=1}^{\infty}\frac{4cl^2 n\pi x}{n^3\pi^3 a}\sin\frac{n\pi}{2}\cdot\sin\frac{n\pi x}{1}\cdot\sin\frac{n\pi\, at}{1}$$

$$\text{i.e., } y(x,t) = \frac{4cl^2}{\pi^3 a}\sum_{n=1}^{\infty}\frac{1}{n^3}\sin\frac{n\pi}{2}\cdot\sin\frac{n\pi x}{1}\cdot\sin\frac{n\pi at}{1}.$$

2.4.2 The Heat Equation and its Solution

A partial differential diffusion equation of the form,

$$\frac{\partial U}{\partial t} = k\nabla^2 U. \qquad ...(1)$$

Physically, the equation commonly arises in situation where k is the thermal diffusivity and U is the temperature. The one-dimensional heat conduction equation is,

$$\frac{\partial U}{\partial t} = \kappa\frac{\partial^2 U}{\partial x^2}. \qquad ...(2)$$

This can be solved by separation of variables using,

$$U(x,t) = X(x)T(t) \qquad ...(3)$$

Then,

$$X\frac{dT}{dt} = \kappa T\frac{d^2X}{dx^2} \quad \text{...(4)}$$

Dividing both sides by kXT gives,

$$\frac{1}{\kappa T}\frac{dT}{dt} = \frac{1}{X}\frac{d^2X}{dx^2} = -\frac{1}{\lambda^2}, \quad \text{...(5)}$$

Where each side must be equal to a constant anticipating the exponential solution in T.

We have picked a negative separation constant so that the solution remains finite at all times and λ has units of length. The solution is,

$$T(t) = Ae^{-\kappa t/\lambda^2}, \quad \text{...(6)}$$

And the X solution is,

$$X(x) = B\cos\left(\frac{x}{\lambda}\right) + C\sin\left(\frac{x}{\lambda}\right) \quad \text{...(7)}$$

The general solution is then,

$$U(x,t) = T(t)X(x) \quad \text{...(8)}$$

$$= Ae^{-\kappa t/\lambda^2}\left[B\cos\left(\frac{x}{\lambda}\right) + C\sin\left(\frac{x}{\lambda}\right)\right] \quad \text{...(9)}$$

$$= e^{-\kappa t/\lambda^2}\left[D\cos\left(\frac{x}{\lambda}\right) + E\sin\left(\frac{x}{\lambda}\right)\right] \quad \text{...(10)}$$

If we are given the boundary conditions,

$$U(0,t) = 0 \quad \text{...(11)}$$

And,

$$U(L,t) = 0 \quad \text{...(12)}$$

Then applying (11) to (10) gives,

$$D\cos\left(\frac{x}{\lambda}\right) = 0 \Rightarrow D = 0, \quad \text{...(13)}$$

And applying (12) to (10) gives,

$$E \sin\left(\frac{L}{\lambda}\right) = 0 \Rightarrow \frac{L}{\lambda} = n\pi \Rightarrow \lambda = \frac{L}{n\pi}, \qquad ...(14)$$

So (10) becomes,

$$U_n(x,t) = E_n e^{-k(n\pi/L)^2 t} \sin\left(\frac{n\pi x}{L}\right) \qquad ...(15)$$

Since the general solution can have any n,

$$U(x,t) = \sum_{n=1}^{\infty} c_n \sin\left(\frac{n\pi x}{L}\right) e^{-\kappa(n\pi/L)^2 t} \qquad ...(16)$$

Now, if we are given an initial condition U (x, 0), we have,

$$U(x,0) = \sum_{n=1}^{\infty} c_n \sin\left(\frac{n\pi x}{L}\right) \qquad ...(17)$$

Multiplying both sides by sin (m π x/L) and integrating from 0 to L gives,

$$\int_0^L \sin\left(\frac{m\pi x}{L}\right) U(x,0)\,dx = \int_0^L \sum_{n=1}^{\infty} c_n \sin\left(\frac{m\pi x}{L}\right) \sin\left(\frac{n\pi x}{L}\right) dx \qquad ...(18)$$

Using the orthogonality of sin (nx) and sin (mx),

$$\sum_{n=1}^{\infty} c_n \int_0^L \sin\left(\frac{n\pi x}{L}\right) \sin\left(\frac{m\pi x}{L}\right) dx \sum_{n=1}^{\infty} \frac{1}{2} L \delta_{mn} C_n \qquad ...(19)$$

$$= \frac{L}{2} c_m \qquad ...(20)$$

$$= \int_0^L \sin\left(\frac{m\pi x}{L}\right) U(x,0)\,dx, \qquad ...(21)$$

So,

$$c_m = \frac{2}{L} \int_0^L \sin\left(\frac{m\pi x}{L}\right) U(x,0)\,dx \qquad ...(22)$$

If the boundary conditions are replaced by the requirement that the derivative of the temperature be zero at the edges,

$$\left.\frac{\partial U}{\partial x}\right|_{(ot)} = 0 \qquad ...(23)$$

$$\left.\frac{\partial U}{\partial x}\right|_{(L,t)} = 0 \qquad ...(24)$$

Following the same procedure as before, a similar answer is found, but with sine replaced by cosine,

$$U(x,t) = \frac{1}{2}c_o + \sum_{n=1}^{\infty} c_n \cos\left(\frac{n\pi x}{L}\right) e^{-\kappa(n\pi/L)^2 t}, \qquad ...(25)$$

Where,

$$c_n = \frac{2}{L}\int_0^L \cos\left(\frac{m\pi x}{L}\right) U(x,0)\,dx \qquad ...(26)$$

One Dimensional Heat Equation with the Initial and Boundary Conditions

The one dimensional heat equation is given by,

$$\frac{\partial u}{\partial t} = a^2 \frac{\partial^2 u}{\partial x^2}$$

The boundary conditions are,

$$\left.\begin{array}{l}(i)\, u(0,t) = k_1^o C \\ (ii)\, u(1,t) = k_2^o C\end{array}\right\} \begin{array}{l}\text{for all } 30 \\ \text{l being the length (rod)}\end{array}$$

The initial condition is,

$$(iii)\, u(x,0) = f(x) \qquad \therefore 0 < x < l$$

Possible Solution of the One Dimensional Heat Equation

The possible solution of the one dimensional heat equation,

$$\frac{\partial u}{\partial t} = a^2 \frac{\partial^2 u}{\partial x^2}$$

- $u(x,t) = \left(A_1\, e^{px} + A_2 e^{-px}\right)e^{a^2p^2t}$.
- $u(x,t) = \left(A_3 \cos px + A_4 \sin px\right)e^{e-a^2p^2t}$.
- $u(x,t) = \left(A_5 x + A_6\right)$.

Problems

1. The ends A and B of a rod x c.m long have their temperature kept at 30°C and 80°C, until steady state conditions prevail. The temperature of the end B is suddenly reduced to 60°C and that of A is increased to 40°C. Let us determine the temperature distribution in the rod after time t.

Solution:

Given:

The ends A and B of a rod x c.m long have their temperature kept at 30°C and 80°C, until steady state conditions prevail. The temperature of the end B is suddenly reduced to 60°C and that of A is increased to 40°C.

If u(x, t) be the temperature distribution at any distance 'x' at any time 't', then u(x, t) satisfies one dimensional heat equation,

$$\frac{\partial u}{\partial t}=a^2\frac{\partial^2 u}{\partial x^2}$$

In steady state condition, the solution is,

$$u(\dot{x})=\frac{50x}{l}+30$$

$$u(0,t)=40, \text{for all } t\geq 0$$

$$u(1,t)=60, \text{for all } t\geq 0$$

$$u(x,0)=\frac{50\,x}{l}+30, \qquad 0\leq x\leq l$$

Let,

$$u(x,t)=u_s(x)+u_t(x,t)$$

In steady state,

$$u_s(x)=\frac{20x}{l}+40$$

In transient state, the conditions are,

- $u_t(0,t)=0$
- $u_t(f,t)=0$

- $u_t(x) = \frac{30x}{l} - 10$

The most suitable solution is,

$$u_t(x,t) = e^{-\alpha^2 p^2 t}(A\cos px + B\sin px) \qquad ...(1)$$

Substituting the first two boundary condition in equation (1), we get,

$$u_t(x,t) = \sum_{n=1}^{\infty} b_n e^{-\alpha^2 p^2 t} \sin px$$

Where,

$$p = \frac{n\pi}{l}$$

Substituting the third boundary condition in the above equation,

$$b_n = \frac{2}{\pi}\int_0^{\pi} u_t(x,0)\sin\frac{n\pi}{l}x\,dx$$

$$= \frac{2}{\pi}\int_0^{\pi}\left(\frac{30x}{l} - 10\right)\sin\frac{n\pi}{l}x\,dx$$

$$b_n = -\frac{20}{n\pi}\left[1 + 2(-1)^n\right]$$

Hence the temperature is,

$$u(x,t) = \frac{20x}{l} + 40 - \frac{20}{\pi}\sum_{n=1}^{\infty}\frac{1}{n}\left(1 + 2(-1)^n\right)\sin\frac{n\pi}{l}xe^{-\frac{n^2\pi^2\alpha^2}{l}t}$$

2. A rod, 30 cm. long, has its ends A and B kept at 20°C and 80°C respectively, until steady state conditions prevail. The temperature at each end is then suddenly reduced to 0°C and kept so. Let us determine the resulting temperature function u(x, t) taking x = 0 at A.

Solution:

Given:

A rod, 30 cm. long, has its ends A and B kept at 20°C and 80°C respectively, until steady state conditions prevail. The temperature at each end is then suddenly reduced to 0°C and kept so.

The temperature function u(x, t) is the solution of the one dimensional heat equation,

$$\frac{\partial u}{\partial t} = \alpha^2 \frac{\partial^2 u}{\partial x^2}$$

When the steady state condition prevails $\frac{\partial u}{\partial t} = 0$ and hence we get $\frac{\partial^2 u}{\partial x^2}$ solving u(x) = ax + b.

When $x = 0\,;\ u = 20 \Rightarrow u(0) = b = 20$

When $x = l\,;\ u = 80 \Rightarrow u(l) = al + 20 = 80$

$$\Rightarrow \frac{60}{l} = a$$

Then,

$$u(x,0) = \frac{60x}{l} + 20$$

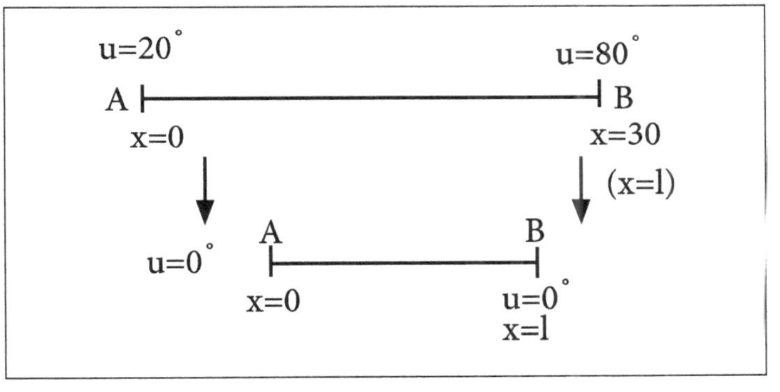

Boundary Conditions

Let us consider length = 30 = l,

$$u(0,t) = 0\,;\, t \ge 0$$

$$u(l,t) = 0\,;\, t > 0$$

$$u(x,0) = \frac{60x}{l} + 20\ ;\ 0 \le x \le l$$

Now the suitable solution which satisfies our boundary conditions is given by,

$$u(x,t) = (A\cos px + B\sin px)\ e^{-\alpha^2 p^2 t} \quad ...(1)$$

Applying condition (i) in (1), put x = 0

$$u(0,t) = A\ e^{-\alpha^2 p^2 t} = 0$$

$$\Rightarrow A = 0 \text{ Sub in } (1)$$

$$u(x,t) = B \sin px\ e^{-\alpha^2 p^2 t} \qquad ...(2)$$

Applying (ii) B.C in(2) i.e., put x = l

$$u(l,\ t) = B \sin pl\ e^{-a^2 p^2 t = 0}$$

$$\Rightarrow \sin pl = 0$$

$$\Rightarrow pl = n\pi$$

$$\text{i.e., } p = \frac{n\pi}{1} \text{sub in} (2)$$

$$u(x,t) = B \sin \frac{n\pi x}{1} e^{-\frac{\alpha^2 n^2 \pi^2 t}{1}} \qquad ...(3)$$

The most general solution is,

$$u(x,t) = \sum_{n=1}^{\infty} B_n \sin \frac{n\pi x}{1} e^{-\frac{a^2 n^2 \pi^2 t}{1^2}} \qquad ...(A)$$

Applying (iii) B.C; put t = 0 in (A)

$$f(x)_{say} = u(x,0) = \sum_{n=1}^{\infty} B_n \sin \frac{n\pi x}{l} \qquad ...(4)$$

We know that half range Fourier sine series for f(x) in (0, 1) is,

$$f(x) = \sum_{n=1}^{\infty} b_n \sin \frac{n\pi x}{l} \qquad ...(5)$$

Where,

$$b_n = \frac{2}{l} \int_0^1 f(x) \sin \frac{n\pi x}{l} dx \qquad ...(6)$$

From equation (4) and (5), we observe that,

$$B_n = b_n = \frac{2}{l} \int_0^1 f(x) \sin \frac{n\pi x}{l} dx$$

$$=\frac{2}{l}\int_0^l\left(\frac{60x}{l}+20\right)\sin\frac{n\pi x}{l}dx$$

$$=\frac{2}{l}\left[\left(\frac{60x}{l}+20\right)\left(\frac{-l}{n\pi}\cos\frac{n\pi x}{l}\right)-\left(\frac{60}{l}\right)\left(\frac{-l^2}{n^2\pi^2}\sin\frac{n\pi x}{l}\right)\right]_0^l$$

$$=\frac{2}{l}\left[\left(\frac{-80l}{n\pi}(-1)^n\right)-\left(\frac{-20l}{nip}\right)\right]$$

$$=\left(\frac{2}{l}\right)\left(\frac{20l}{n\pi}\right)\left[1-4(-1)^n\right]$$

$$B_n=\frac{40}{n\pi}\left[1-4(-1)^n\right]\text{Sub in (A)}$$

$$u(x,t)=\sum_{n=1}^{\infty}\frac{40}{n\pi}\left(1-4(-1)^n\right)\sin\frac{n\pi x}{l}e^{-\frac{\alpha^2n^2\pi^2t}{l^2}}$$

Replace l by 30 cm,

$$u(x,t)=\frac{40}{\pi}\sum_{n=1}^{\infty}\frac{1}{n}\left[1-4(-1)^n\right]\sin\frac{n\pi x}{30}e^{-\frac{\alpha^2n^2\pi^2t}{900}}$$

2.4.3 Two Dimensional Wave Equation and its Solution

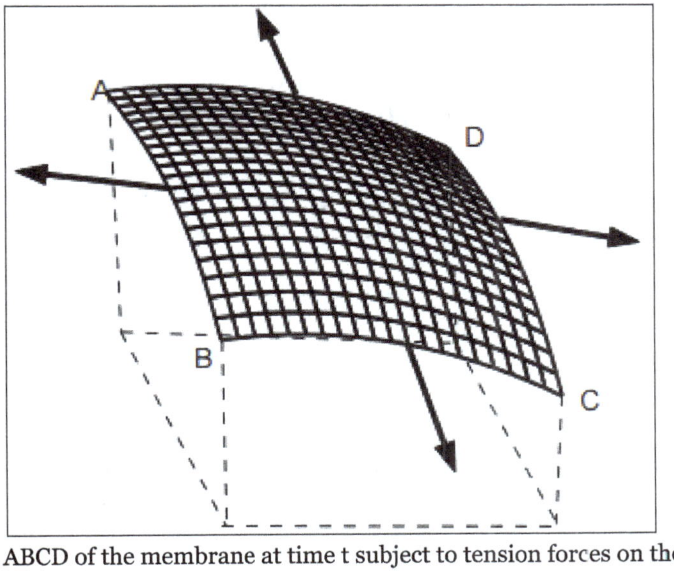

Element ABCD of the membrane at time t subject to tension forces on the edges.

Consider a homogeneous, well stretched, flexible membrane whose boundary is held fixed in the (x, y) plane. As with the case of the string, we assume that the tension force is the dominant force governing the vibrations. We also assume that the vibrations are small and vertical when compared to the size of the membrane. Our aim is to construct a model for the vertical displacement function.

Let u(x, y, t) represent the vertical displacement at time t at the point (x, y) of the membrane. Consider a small rectangular element of the membrane, ABCD, that correspond to the points (x, y), (x + Δx, y), (x + Δx, y + Δy) and (x, y + Δy). We assume that this portion of the membrane has a vertical motion under the action of the tension forces. As in the case of the string, the flexibility of the membrane implies that the tension $\vec{T}$ (x, y, t) has constant magnitude T at each point of the membrane and that $\vec{T}$ is tangent to the surface.

Since we are dealing with small vibrations, we can assume that at each time t and at each point, the angle θ between the normal to the surface and the z-axis is close to 0. By using linear approximation, sin θ ≈ tan θ, similar arguments as those used for the string, imply that the total vertical force F acting on the element ABCD can be approximated as follows:

$$F \approx T\left[\Delta y\left(\frac{\partial u}{\partial x}(x+\Delta x,y,t)-\frac{\partial u}{\partial x}(x,y,t)\right)+\Delta x\left(\frac{\partial u}{\partial y}(x,y+\Delta y,t)-\frac{\partial u}{\partial y}(x,y,t)\right)\right] \quad ...(1)$$

We can also use Newton's second law to express F as,

$$F \approx \Delta m\frac{\partial^2 u}{\partial t^2}(x,y,t) \approx p\Delta x\ \Delta y\frac{\partial^2 u}{\partial t^2}(x,y,t) \qquad ...(2)$$

Where, $\Delta m \approx \rho \Delta x\, \Delta y$. It denotes the mass of the rectangular element ABCD and with ρ it represents the surface mass density of the membrane. After substituting equation (1), the expression for F given in equation (2) and dividing by ΔxΔy, we get,

$$\frac{\partial^2 u}{\partial t^2}(x,y,t) \approx T\left[\frac{\frac{\partial u}{\partial x}(x+\Delta x,y,t)-\frac{\partial u}{\partial x}(x,y,t)}{\Delta x}+\frac{\frac{\partial u}{\partial y}(x,y+\Delta y,t)-\frac{\partial u}{\partial y}(x,y,t)}{\Delta y}\right] \quad ...(3)$$

The approximation in equation (3) becomes better and better, as Δx and Δy become smaller and smaller.

At the limit, we obtain the two dimensional wave equation as,

$$\frac{\partial^2 u}{\partial t^2}(x,y,t)=c^2\left(\frac{\partial^2 u}{\partial x^2}(x,y,t)+\frac{\partial^2 u}{\partial y^2}(x,y,t)\right)$$

Where, $c^2 = T/\rho$ is the wave's speed.

2.5 Laplace Equation in Polar

A Laplace's equation in R2 in terms of the usual (i.e., Cartesian) (x, y) coordinate system is,

$$\frac{\partial^2 u}{\partial x^2}+\frac{\partial^2 u}{\partial y^2}=u_{xx}+u_{yy}=0. \qquad ...(1)$$

The Cartesian coordinates can be represented by the polar coordinates as follows,

$$\begin{cases} x=r\cos\theta \\ y=r\sin\theta \end{cases} \qquad ...(2)$$

Let us first compute the partial derivatives of x, y w.r.t. r, θ,

$$\begin{cases} \dfrac{\partial x}{\partial r}=\cos\theta, & \dfrac{\partial x}{\partial \theta}=r\sin\theta; \\ \dfrac{\partial y}{\partial r}=\sin\theta, & \dfrac{\partial y}{\partial \theta}=r\cos\theta. \end{cases} \qquad ...(3)$$

To do so, let's compute $\frac{\partial u}{\partial r}$ first. We will use the Chain Rule since (x, y) are functions of (r, θ) as shown in (2),

$$\frac{\partial u}{\partial r}=\frac{\partial u}{\partial x}\frac{\partial x}{\partial r}+\frac{\partial u}{\partial y}\frac{\partial y}{\partial r}$$

$$=\frac{\partial u}{\partial x}\cos\theta+\frac{\partial u}{\partial y}\sin\theta \qquad \text{using} \qquad ...(3)$$

$$=\cos\theta\frac{\partial u}{\partial x}+\sin\theta\frac{\partial u}{\partial y} \qquad ...(4)$$

Now, let's compute $\frac{\partial^2 u}{\partial r^2}$. Noticing that both $\frac{\partial u}{\partial x}$ and $\frac{\partial u}{\partial y}$ are functions of (x, y) And using (3), we have,

$$\frac{\partial^2 u}{\partial r^2}=\cos\theta\frac{\partial}{\partial r}\frac{\partial u}{\partial x}+\sin\theta\frac{\partial}{\partial r}\frac{\partial u}{\partial y}$$

$$=\cos\theta\left(\frac{\partial}{\partial x}\frac{\partial u}{\partial x}\frac{\partial x}{\partial r}+\frac{\partial}{\partial y}\frac{\partial u}{\partial x}\frac{\partial y}{\partial r}\right)+\sin\theta\left(\frac{\partial}{\partial x}\frac{\partial u}{\partial y}\frac{\partial x}{\partial r}+\frac{\partial}{\partial y}\frac{\partial u}{\partial y}\frac{\partial y}{\partial r}\right)$$

$$=\cos^2\theta\frac{\partial^2 u}{\partial x^2}+2\cos\theta\sin\theta\frac{\partial^2 u}{\partial x\partial y}+\sin^2\theta\frac{\partial^2 u}{\partial y^2} \qquad ...(5)$$

Similarly, let's compute $\frac{\partial u}{\partial \theta}$ and $\frac{\partial^2 u}{\partial \theta^2}$.

$$\frac{\partial u}{\partial \theta}=\frac{\partial u}{\partial x}\frac{\partial x}{\partial \theta}+\frac{\partial u}{\partial y}\frac{\partial y}{\partial \theta}$$

$$=\frac{\partial u}{\partial x}(-r\sin\theta)+\frac{\partial u}{\partial y}(r\cos\theta)$$

$$=-r\sin\theta\frac{\partial u}{\partial x}+r\cos\theta\frac{\partial u}{\partial y}.$$

$$\frac{\partial^2 u}{\partial \theta^2}=-r\cos\theta\frac{\partial u}{\partial x}-r\sin\theta\frac{\partial}{\partial \theta}\frac{\partial u}{\partial x}-r\sin\theta\frac{\partial u}{\partial y}+r\cos\theta\frac{\partial}{\partial \theta}\frac{\partial u}{\partial y}$$

$$=-r\cos\theta\frac{\partial u}{\partial x}-r\sin\theta\left(\frac{\partial}{\partial x}\frac{\partial u}{\partial x}\frac{\partial x}{\partial \theta}+\frac{\partial}{\partial y}\frac{\partial u}{\partial x}\frac{\partial y}{\partial \theta}\right)-r\sin\theta\frac{\partial u}{\partial y}+r\cos\theta\left(\frac{\partial}{\partial x}\frac{\partial u}{\partial y}\frac{\partial x}{\partial \theta}+\frac{\partial}{\partial y}\frac{\partial u}{\partial y}\frac{\partial y}{\partial \theta}\right)$$

$$=-r\cos\theta\frac{\partial u}{\partial x}-r\sin\theta\left(\frac{\partial^2 u}{\partial x^2}(-r\sin\theta)+\frac{\partial^2 u}{\partial x\partial y}r\cos\theta\right)$$

$$-r\sin\theta\frac{\partial u}{\partial x}+r\cos\theta\left(\frac{\partial^2 u}{\partial x^2}(-r\sin\theta)+\frac{\partial^2 u}{\partial y^2}r\cos\theta\right)$$

$$=-r\left(\cos\theta\frac{\partial u}{\partial x}+r\sin\theta\frac{\partial u}{\partial y}\right)+r^2\left(\sin^2\theta\frac{\partial^2 u}{\partial x^2}-2\cos\theta\sin\theta\frac{\partial^2 u}{\partial x\partial y}+\cos^2\theta\frac{\partial^2 u}{\partial y^2}\right)$$

Dividing both sides by r2 and using (4), we have,

$$\frac{1}{r^2}\frac{\partial^2 u}{\partial \theta^2}=-\frac{1}{r}\frac{\partial u}{\partial r}+\sin^2\theta\frac{\partial^2 u}{\partial x^2}-2\cos\theta\sin\theta\frac{\partial^2 u}{\partial x\partial y}+\cos^2\theta\frac{\partial^2 u}{\partial y^2} \qquad ...(6)$$

Finally, adding (5) and (6), using the obvious relation cos2θ +sin2θ = 1, we have,

$$\frac{\partial^2 u}{\partial r^2}+\frac{1}{r^2}\frac{\partial^2 u}{\partial \theta^2}=-\frac{1}{r}\frac{\partial u}{\partial r}+\frac{\partial^2 u}{\partial x^2}+\frac{\partial^2 u}{\partial y^2},$$

Which can be cleaned up as,

$$\frac{\partial^2 u}{\partial x^2}+\frac{\partial^2 u}{\partial y^2}=\frac{\partial^2 u}{\partial r^2}+\frac{1}{r}\frac{\partial u}{\partial r}+\frac{1}{r^2}\frac{\partial^2 u}{\partial \theta^2}$$

Hence, Laplace's equation (1) becomes,

$$u_{xx} + u_{yy} = u_{rr} + \frac{1}{r}u_r + \frac{1}{r^2}u_{\theta\theta} = 0.$$

Once we derive Laplace's equation in the polar coordinate system, it is easy to represent the heat and wave equations in the polar coordinate system. For the heat equation, the solution u(x, y, t) = u(r, θ, t) satisfies,

$$u_t = k\left(u_{xx} + u_{yy}\right) = k\left(u_{rr} + \frac{1}{r}u_r + \frac{1}{r^2}u_{\theta\theta}\right), \; k > 0, \text{ diffusivity},$$

whereas for the wave equation, we have,

$$u_t = c^2\left(u_{xx} + u_{yy}\right) = c^2\left(u_{rr} + \frac{1}{r}u_r + \frac{1}{r^2}u_{\theta\theta}\right), \; c > 0;$$

2.5.1 Cylindrical and Spherical Coordinates

The first step in solving a boundary value problem is generally the introduction of coordinates in which the boundary surface S has a simple representation. Cylindrical symmetry (a cylinder as a region T) calls for cylindrical coordinates r, θ, z related to x, y, z by,

$$x = r\cos\theta, \qquad y = r\sin\theta, \qquad z = z$$

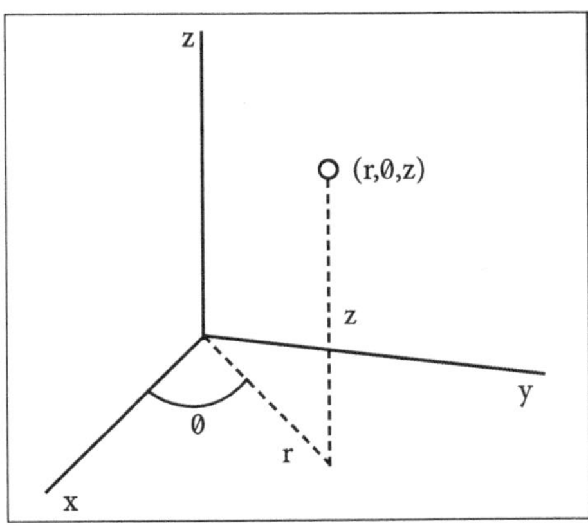

Cylindrical coordinates.

$$\left(r \geqq 0, \; 0 \leqq \theta \leqq 2\pi\right)$$

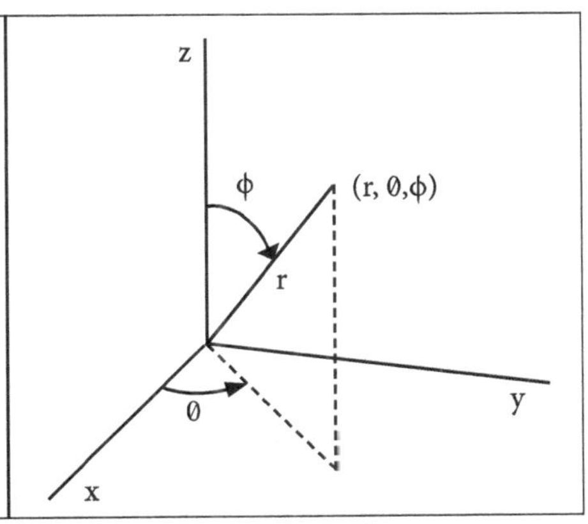

Spherical coordinates.

$$\left(r \geqq 0, \; 0 \leqq \theta \leqq 2\pi, 0 \leqq \phi \leqq \pi\right)$$

For these we get $\nabla^2 u$ immediately by adding u_{zz}.

$$\nabla^2 u = \frac{\partial^2 u}{\partial r^2} + \frac{1}{r}\frac{\partial u}{\partial r} + \frac{1}{r^2}\frac{\partial^2 u}{\partial \theta^2} + \frac{\partial^2 u}{\partial z^2}$$

Laplacian in Spherical Coordinates

Spherical symmetry (a ball as region T bounded by a sphere S) requires spherical.

Co-ordinates r, θ, Φ related to x, y, z, by,

$$x = r\cos\theta\sin\phi, \quad y = r\sin\theta\sin\phi, \quad z = r\cos\phi$$

Using the chain rule $\nabla^2 u$ we obtain in spherical coordinates,

$$\nabla^2 u = \frac{\partial^2 u}{\partial r^2} + \frac{2}{r}\frac{\partial u}{\partial r} + \frac{1}{r^2}\frac{\partial^2 u}{\partial \phi^2} + \frac{\cot\phi}{r^2}\frac{\partial u}{\partial \phi} + \frac{1}{r^2\sin^2\phi}\frac{\partial^2 u}{\partial \theta^2}$$

We leave the details as an exercise. It is sometimes practical to write in the form,

$$\nabla^2 u = \frac{1}{r^2}\left[\frac{\partial}{\partial r} + \left(r^2\frac{\partial u}{\partial r}\right) + \frac{1}{\sin\phi}\frac{\partial}{\partial \phi} + \left(\sin\phi\frac{\partial u}{\partial \phi}\right) + \frac{1}{\sin^2\phi}\frac{\partial^2 u}{\partial \theta^2}\right]$$

Use of Fourier-Legendre Series

Interior Problem: Potential Within the Sphere S. We consider a series of terms from,

$$u(r,\phi) = \sum_{n=0}^{\infty} A_n r^n P_n(\cos\phi) \qquad ...(1)$$

Since Sis given by for (1) to satisfy the Dirichlet condition on the sphere S, we must have,

$$u(R,\phi) = \sum_{n=0}^{\infty} A_n R^n P_n(\cos\phi) = f(\phi); \qquad ...(2)$$

That is, (2) must be the Fourier-Legendre series of given by,

$$A_n R^n = \frac{2n+1}{2}\int_{-1}^{1} \tilde{f}(w)P_n(w)dw \qquad ...(3)$$

where f(ϕ) denotes f' (ϕ) as a function of w = cosϕ. Since dw = −sinϕ dϕ and the limits of integration -1 and 1 correspond to ϕ = π and ϕ = 0 respectively, we also obtain

$$A_n R^n = \frac{2n+1}{2}\int_{0}^{\pi} f(\phi)P_n(\cos\phi)\sin\phi\, d\phi,$$

If f(ϕ)and f' (ϕ) are piecewise continuous on the interval $0 \leqq \phi \leqq \pi$ then the series (1) with coefficients. solves our problem for points inside the sphere because it can be shown that under these continuity assumptions the with coefficients (3) gives the derivatives occurring by term wise differentiation, thus justifying in Derivation.

Exterior Problem

Potential Outside the Sphere S. Outside the sphere we cannot use the functions. Proceeding as before leads to the solution of the exterior problem,

$$u(r,\phi)=\sum_{n=0}^{\infty}\frac{B_n}{r^{n+1}}P_n(\cos\phi)$$

$$B_n=\frac{2n+1}{2}R^{n+1}\int_0^{\pi}f(\phi)P_n(\cos\phi)\sin\phi\, d\phi.$$

The next example illustrates all this for a sphere of radius 1 consisting of two hemispheres that are separated by a small strip of insulating material along the equator, so that these hemispheres can be kept at different potentials (110 V and 0 V).

3

Complex Analysis and Complex Integration

3.1 Complex Analysis: Analytic Function

Continuous

A function f(z) is said to be continuous at z= z_0 if f(zo) is defined and $\lim_{z \to z_0} f(z) = f(z_0)$

Derivative of f(z)

A function f (z) is said to be differentiable at a fixed point z if $\lim_{\Delta z \to z_0} \frac{f(z+\Delta z)-f(z)}{\Delta z}$ exists.

This limit is called the derivative of f (z) at the point z and it is denoted by f (z).

Note: Here z + Δz is a neighbourhood point of z with small distance Δz, can be chosen in any side of z. This is an important aspect in the above definition.

Analytic Function

A function f (z) is said to be analytic at a point z = a in a region R if:

- f (z) is differentiable at z = a.
- f (z) is differentiable at all points for some neighbourhood of z = a.

A function f (z) is said to be analytic in a region R if f (z) is analytic at all points in the region R.

Note: Instead of the term analytic in R, the terms holomorphic in R and regular in R are also used.

Analytic or Holomorphic/Regular Function

A function is said to be analytic at a point if its derivative exists not only at the point but also in some neighbourhood of that point.

Entire Function

A function which is analytic everywhere in the finite plane is called an entire function.

Example: e^z, sinZ, cosz, sinhz, coshz.

Problems

1. Let us prove that $u = e^{-y} \cos x$ and $v = e^{-x} \sin y$ satisfy Laplace equations, but that u + iv is not an analytic function of z.

Solution:

Given:

$$u = e^{-y} \cos x$$

$$v = e^{-x} \sin y$$

$$u = e^{-y} \cos x$$

$$u_x = -e^{-y} \sin x$$

$$u_{xx} = -e^{-y} \cos x.$$

$$u_y = -e^{-y} \cos x$$

$$u_{yy} = e^{-y} \cos x$$

$$\therefore u_{xx} + u_{yy} = \left(-e^{-y} \cos x\right) + \left(e^{-y} \cos x\right) = 0$$

$\Rightarrow$ u is harmonic, u satisfies Laplace equation.

$$v = e^{-x} \sin y$$

$$v_x = -e^{-x} \sin y v_y = e^{-x} \cos y$$

$$v_{xx} = e^{-x} \sin y v_{yy} = -e^{-x} \sin y$$

$$\therefore v_{xx} + v_{yy} = e^{-x} \sin y - e^{-x} \sin y = 0$$

$\Rightarrow$ v is harmonic equation.

Also,

$$u_x \neq v_y \text{ and } u_y \neq -v_x.$$

$\therefore\ u+iv$ is not analytic function.

2. Here we shall prove that every analytic function w = u + iv can be expressed as a function of z alone, not as a function of $\bar{z}$.

Solution:

Let $z=x+iy$ and $\bar{z}=x-iy$ and $w=u(x,y)+v(x,y)$,

$$\therefore x=\frac{z+\bar{z}}{2} \text{ and } y=\frac{z-\bar{z}}{2i}.$$

Hence u and v and also w may be considered as a function of z and $\bar{z}$.

Consider $\frac{\partial w}{\partial \bar{z}}=\frac{\partial u}{\partial \bar{z}}+1\frac{\partial v}{\partial \bar{z}}$

$$=\left(\frac{\partial u}{\partial x}.\frac{\partial x}{\partial \bar{z}}+\frac{\partial u}{\partial y}\frac{\partial y}{\partial \bar{z}}\right)+i\left(\frac{\partial v}{\partial x}\frac{\partial x}{\partial \bar{z}}+\frac{\partial v}{\partial y}\frac{\partial y}{\partial \bar{z}}\right)$$

$$=\left(u_x\left(\frac{1}{2}\right)+u_y\left(\frac{-1}{2i}\right)\right)+i\left(v_x\left(\frac{1}{2}\right)+v_y\left(-\frac{1}{2i}\right)\right)$$

$$=\frac{u_x}{2}-\frac{u_y}{2i}+\frac{iv_x}{2}-\frac{v_y}{2}.$$

Since w is analytic function $u_x=v_y$ and $u_y=-v_x$

$$\frac{\partial w}{\partial \bar{z}}=\frac{u_x}{2}+\frac{iu_y}{2}-\frac{iu_y}{2}-\frac{u_x}{2}$$

$$=0$$

$\therefore$w is independent of $\bar{z}$.

Hence w is a function of z alone.

3. If f(z) is an analytic function of z, Let us prove that $\left(\frac{\partial^2}{\partial x^2}+\frac{\partial^2}{\partial y^2}\right)\log|f(z)|=0$.

Solution:

Given:

f(z) is analytic function.

Let,

f(z) = u + iv

Since f(z) is analytic function we've $u_x = v_y$ and $u_y = -v_x$.

To prove:

$$\left(\frac{\partial^2}{\partial x^2}+\frac{\partial^2}{\partial y^2}\right)\log|f(z)|=0$$

$$f(z)=u+iv \qquad |f(z)=|u+iv|=\sqrt{u^2+v^2}$$

$$\log|f(z)|=\log\sqrt{u^2+v^2}=\frac{1}{2}\log\left(u^2+v^2\right)$$

$$\text{L.H.S}=\left(\frac{\partial^2}{\partial x^2}+\frac{\partial^2}{\partial y^2}\right)\log|f(z)|$$

$$=\frac{\partial^2}{\partial x^2}\log|f(z)|+\frac{\partial^2}{\partial y^2}\log|f(z)|$$

$$=\frac{\partial}{\partial x}\log|f(z)|=\frac{\partial}{\partial x}\frac{1}{2}\log\left(u^2+v^2\right)=\frac{1}{2}\left[\frac{\partial}{\partial x}\log\left(u^2+v^2\right)\right]$$

$$=\frac{1}{2}\left[\frac{1}{u^2+v^2}\left(2uu_x+2vv_x\right)\right]$$

$$\frac{\partial}{\partial x}\log|f(z)|=\frac{uu_x+vv_x}{u^2+v^2}.$$

$$\frac{\partial^2}{\partial x^2}\log|f(z)|=\frac{\partial}{\partial x}\left[\frac{\partial}{\partial x}\log|f(z)|\right]=\frac{\partial}{\partial x}\left[\frac{uu_x+vv_x}{u^2+v^2}\right]$$

$$=\frac{1}{\left(u^2+v^2\right)^2}\left[u\left(u^2+v^2\right)u_{xx}+\left(u^2+v^2\right)v_{xx}+\left(u^2+v^2\right)\left(u_x^2+v_x^2\right)-2\left(uu_x+vv_x\right)^2\right].$$

$$=\frac{1}{\left(u^2+v^2\right)^2}\left[u\left(u^2+v^2\right)u_{xx}+v\left(u^2+v^2\right)v_{xx}+\left(u^2+v^2\right)\left(u_x^2+v_x^2\right)\right.$$

$$\left.-2\left(u^2u_x^2+v^2v_x^2+2uvu_xv_x\right)\right] \qquad \ldots(1)$$

Since f(z) is analytic $u_{xx}+u_{yy}=0$ and $v_{xx}+v_{yy}=0$, also $u_x=v_y$ and $u_y=-v_x$.

Similarly $\frac{\partial^2}{\partial y^2}\log|f(z)|$

$$=\frac{1}{\left(u^2+v^2\right)^2}\left[u\left(u^2+v^2\right)u_{yy}+v\left(u^2+v^2\right)v_{yy}\right.$$

$$+\left(u^2+v^2\right)\left(u_y^2+v_y^2\right)-2\left(u^2u_y^2+v^2v_y^2+2uvu_yv_y\right)\Big] \qquad ...(2)$$

(1) + (2) gives.

$$\left(\frac{\partial^2}{\partial x^2}+\frac{\partial^2}{\partial y^2}\right)\log|f(z)|=\frac{}{\left(u^2 \quad v^2\right)^2}\Big[u\left(u^2+v^2\right)\left(u_{xx}+u_{yy}\right)+v\left(u^2+v^2\right)\left(u_{xx}+u_{yy}\right)$$

$$+\left(u^2+v^2\right)\left(u_x^2+v_x^2\right)$$

$$+\left(u^2+v^2\right)\left(u_y^2+v_y^2\right)-2u^2u_x^2-2u^2u_y^2-2v^2v_x^2-2v^2v_y^2$$

$$-4uv\,u_xv_x-4uv\,u_yv_y\Big]$$

$$=\frac{}{\left(u^2 \quad v^2\right)}\Big[\left(u^2+v^2\right)\left(u_x^2+u_y^2\right)+\left(u^2+v^2\right)\left(u_y^2+u_x^2\right)\Big]$$

$$-2u^2\left(u_x^2+u_y^2\right)-2v^2\left(v_x^2+v_y^2\right)-4uvu_x\left(-u_y\right)-4uvu_xu_y\Big].$$

$$=\frac{1}{\left(u^2+v^2\right)^2}\Big[2\left(u^2+v^2\right)\left(u_x^2+u_y^2\right)-2u^2\left(u_x^2+u_y^2\right)-2v^2\left(u_x^2+u_y^2\right)\Big]$$

$$=\frac{1}{\left(u^2+v^2\right)^2}\Big[2\left(u^2+v^2\right)\left(u_x^2+u_y^2\right)-2\left(u_x^2+u_y^2\right)\left(u^2+v^2\right)\Big]$$

$$=0$$

$$\therefore\left(\frac{\partial^2}{\partial x^2}+\frac{\partial^2}{\partial y^2}\right)\log|f(z)|=0.$$

Hence proved.

We shall now obtain the necessary conditions for a complex function $f(z) = u(x,y) + iv(x,y)$ to be analytic.

Let f(z) = u + iv be analytic in a region R in z plane. So, f(z) is differentiable at R.

$$\Rightarrow f'(z)=\lim_{\Delta z\to 0}\frac{f(z+\Delta z)-f(z)}{\Delta z} \text{ exists in R} \qquad ...(1)$$

Let,

$$z=x+iy$$

$$\Delta z=\Delta x+i\Delta y$$

$$z+\Delta z=x+iy+\Delta x+i\Delta y$$

$$= x + \Delta x + i(y + \Delta y)$$

Now,

$$f(z) = u(x, y) + iv(x, y)$$

$$f(z + \Delta z) = u(x + \Delta x, y + \Delta y) + iv(x + \Delta x, y + \Delta y) \qquad ...(2)$$

Case (i): Choose z + Δz along the horizontal line through z.

Then,

$$\Delta y = 0$$

So,

$$\Delta z = \Delta x$$

$$\therefore \Delta z \to 0 \Rightarrow \Delta x \to 0$$

Now from (1) and (2),

$$\Rightarrow f'(z) = \lim_{\Delta \to 0} \frac{[u(x+\Delta x, y+\Delta y) + iv(x+\Delta x, y+\Delta y)] - [u(x,y) + iv(x,y)]}{\Delta x + i\Delta y}$$

$$= \lim_{\Delta x \to 0} \frac{u(x+\Delta x, y) + iv(x+\Delta x, y) - u(x,y) - iv(x,y)}{\Delta x}$$

$$= \lim_{\Delta x \to 0} \frac{[u(x+\Delta x, y) - u(x,y)] + i[v(x+\Delta x, y) - v(x,y)]}{\Delta x}$$

$$= \lim_{\Delta x \to 0} \frac{u(x+\Delta x, y) - u(x,y)}{\Delta x} + i \quad \lim_{\Delta x \to 0} \frac{v(x+\Delta x, y) - v(x,y)}{\Delta x}$$

$$f'(z) = \frac{du}{dx} + i\frac{dv}{dx} \qquad ...(3)$$

Case (ii): Choose z + Δz along the vertical line through z.

Then,

$$\Delta x = 0.$$

$$\ddot{A}z = i\ddot{A}y$$

$\therefore \text{Äz} \to 0 \text{ gives Äy} \to 0$

From (1) and (2),

$$\Rightarrow f'(z) = \lim_{\Delta z \to 0} \frac{\left[u(x+\Delta x, y+\Delta y) + iv(x+\Delta x, y+\Delta y)\right] - \left[u(x, y) + iv(x, y)\right]}{\Delta x + i\Delta y}$$

$$= \lim_{\Delta y \to 0} \frac{u(x,y+\Delta y) + iv(x,y+\Delta y) - u(x,y) - iv(x,y)}{i\Delta y}$$

$$= \frac{1}{i} \lim_{\Delta y \to 0} \frac{\left[u(x, y+\Delta y) - u(x,y)\right] + i\left[v(x,y+\Delta y) - v(x,y)\right]}{\Delta y}$$

$$= \frac{1}{i} \lim_{\Delta y \to 0} \frac{u(x,y+\Delta y) - u(x,y)}{\Delta y} + \lim_{\Delta y \to 0} \frac{v(x,y+\Delta y) - v(x,y)}{\Delta y}$$

$$= \frac{1}{i}\frac{du}{dy} + i\frac{dv}{dy}$$

$$f'(z) = -i\frac{du}{dy} + i\frac{dv}{dy} \qquad \ldots(4)$$

From (3) and (4) we get,

$$\frac{du}{dx} + i\frac{dv}{dx} = -i\frac{du}{dy} + i\frac{dv}{dy}$$

Equating the real and imaginary parts, we get,

$$\frac{du}{dx} = \frac{dv}{dy} = \frac{du}{dy} = -\frac{dv}{dx}$$

The above equations are known as Cauchy-Riemann equations (or) C-R equations.

Sufficient Condition for A Function F(Z) to be Analytic

Continuous single valued function f (z) = u + iv is analytic in a region R, if:

- The four partial derivatives $\frac{du}{dx}, \frac{du}{dy}, \frac{dv}{dx}, \frac{dv}{dy}$ exist and all are continuous.

- $\frac{du}{dx} = \frac{dv}{dy} \cdot \frac{du}{dy} = -\frac{dv}{dx}$

Proof:

By Taylor's series,

$$f(x+h, y+k) = f(x,y) + \left(h\frac{d}{dx} + k\frac{d}{dy}\right) f(x,y) + \frac{1}{2!}\left(h\frac{d}{dx} + k\frac{d}{dy}\right)^2 f(x,y) + \ldots \quad \ldots(1)$$

Let,

$$z = x + iy \Rightarrow \Delta z = \Delta x + i\Delta y$$

$$z + \Delta z = u(x+iy) + (\Delta x + i\Delta y) = (x + \Delta x) + i(y + \Delta y)$$

$$f(z + \Delta z) = u(x + \Delta x, y + \Delta y) + iv(x + \Delta x, y + \Delta y)$$

$$u(x,y) = \left(\Delta x\frac{d}{dx} + \Delta y\frac{d}{dy}\right) u(x,y) + \frac{1}{2!}\left(\Delta x\frac{d}{dx} + \Delta y\frac{d}{dy}\right)^2$$

$$u(x,y) + \ldots + i\left[v(x,y) + \left(\Delta x\frac{d}{dx} + \Delta y\frac{d}{dy}\right) v(x,y) + \ldots\right] \{\text{by } (1)\}$$

$$= u(x,y) + \Delta x\frac{du}{dx} + \Delta y\frac{du}{dy} + i\left(v(x,y) + \Delta x\frac{dv}{dx} + \Delta y\frac{dy}{dy}\right) + \ldots$$

$$= u(x,y) + iv(x,y) + \Delta x\left(\frac{du}{dx} + i\frac{dv}{dx}\right) + \Delta y\left(\frac{du}{dx} + i\frac{dv}{dy}\right)$$

[Omitting the higher powers of Δx, Δy],

$$= f(z) + \Delta x\left(\frac{du}{dx} + i\frac{dv}{dx}\right) + \Delta y\left(-\frac{dv}{dx} + i\frac{du}{dx}\right) \text{ [by CR equation]}$$

$$= f(z) + \Delta x\left(\frac{du}{dx} + i\frac{dv}{dx}\right) + \Delta y\left(i^2\frac{dv}{dx} + i\frac{du}{dx}\right) \quad \text{[Since } i^2 = -1\text{]}$$

$$= f(z) + \Delta x\left(\frac{du}{dx} + i\frac{dv}{dx}\right) + i\Delta y\left(\frac{du}{dx} + i\frac{dv}{dx}\right)$$

$$= f(z) + \left(\frac{du}{dx} + i\frac{dv}{dx}\right)(\Delta x + i\Delta y)$$

$$f(z+\Delta z)=f(z)+\left(\frac{du}{dx}+i\frac{dv}{dx}\right)\Delta z$$

$$\Rightarrow f(z+\Delta z)-f(z)=\left(\frac{du}{dx}+i\frac{dv}{dx}\right)\Delta z$$

$$\Rightarrow \frac{f(z+\Delta z)-f(z)}{z}=\frac{du}{dx}+\frac{dv}{dx}$$

$$\Rightarrow \lim_{\Delta z\to o}\frac{f(z+\Delta z)-f(z)}{\Delta z}=\lim_{\Delta z\to o}\frac{du}{dx}+i\frac{dv}{dx}$$

$$\Rightarrow f'(z)=\frac{du}{dx}+i\frac{dv}{dx}=\frac{dv}{dy}-i\frac{du}{dy} \qquad \text{(by C-R equation)}$$

$\Rightarrow f'(z)$ exits and $\frac{du}{dx},\frac{du}{dy},\frac{dv}{dx},\frac{dv}{dy}$ are all continuous.

Polar Form of Cauchy Riemann Equations

Let,

$$z=re^{i\theta} \text{ and } f(z)=p(r,\,\theta)+iQ(r,\,\theta)$$

Then,

$$p(r,\,\theta)+iQ(r,\,\theta)=f\left(re^{i\theta}\right) \qquad ...(1)$$

Differentiating (1) partially w.r.t to r, we get,

$$\frac{\partial P}{\partial r}+i\frac{\partial Q}{\partial r}=f'\left(re^{i\theta}\right)e^{i\theta}$$

Differentiating (1) partially w.r.t to θ, we get,

$$\frac{\partial p}{\partial \theta}+i\frac{\partial Q}{\partial \theta}=f'\left(re^{i\theta}\right).r.e^{i\theta}$$

$$=ri\left|f'\left(re^{i\theta}\right)e^{i\theta}\right|$$

$$=ri\left[\frac{\partial p}{\partial r}+i\frac{\partial Q}{\partial r}\right]$$

$$=ir\frac{\partial p}{\partial r}+i^2r\frac{\partial Q}{\partial r}$$

$$= ir\frac{\partial p}{\partial r} - r\frac{\partial Q}{\partial r}$$

[Since $i^2 = -1$]

Equating the real and imaginary parts, we get,

$$\frac{\partial p}{\partial \theta} = -r\frac{\partial Q}{\partial r}; \quad \frac{\partial Q}{\partial \theta} = r\frac{\partial p}{\partial r}$$

$$\Rightarrow \frac{\partial}{\partial r} = \frac{-1}{r}\frac{\partial p}{\partial \theta}; \quad \frac{\partial p}{\partial r} = \frac{i}{r}\frac{\partial Q}{\partial \theta}$$

The above equations are called Cauchy Riemann equations in polar form.

Problems

1. Let us test whether the function f(z) = z3 + z is analytic or not.

Solution:

Given:

$$f(z) = z^3 + z$$

$$= x^3 + 3x^2(iy) + 3x(iy)^2 + (iy)^3 + x + iy$$

$$\left[\text{since } (a+b)^3 = a^3 + 3a^2b + 3ab^2 + b^3\right]$$

$$= x^3 + i3x^2y - 3xy^2 - iy^3 + x + iy$$

$$\left[\text{Since } i^2 = -1, i^3 = -i\right]$$

$$= \left(x^3 - 3xy^2 + x\right) + i\left(3x^2y - y^3 + y\right)$$

Equating real and imaginary parts, we get,

$$u = x^3 - 3xy^2 + x, \qquad v = 3x^2y - y^3 + y$$

$$\Rightarrow \frac{du}{dx} = 3x^2 - 3y^2 + 1; \qquad \frac{dv}{dx} = 6xy - 0 + 0 = 6xy$$

$$\frac{du}{dy} = 0 - 3x(2y) + 0; \quad \frac{dv}{dy} = 3x^2(1) - 3y^2 + 1$$

$$= -6xy; \qquad = 3x^2 - 3y^2 + 1$$

$$\Rightarrow \frac{du}{dx} = \frac{dv}{dy}; \qquad \frac{du}{dy} = \frac{-dv}{dx}$$

⇒ C-R equations are satisfied.

⇒ f(z) = z3 + z is analytic.

2. We shall verify wether f(z) = sinhz is analytic using C - R equations.

Solution:

Given:

$$f(z) = \sinh z$$

$$\Rightarrow u + iv = \sinh(x + iy)$$

$$= \frac{1}{i}\sin i(x + iy) \qquad [\text{Since} \sin i\theta = i \sinh \theta]$$

$$= \frac{1}{i}\sin(ix + i^2 y)$$

$$= \frac{1}{i}\sin(ix - y) \quad [\text{Since } i^2 = -1]$$

$$= \frac{1}{i}\{\sin ix \cos y - \cos ix \sin y\}$$

$$[\text{Since } \sin(a - b) = \sin a \cos b - \cos a \sin b]$$

$$= \frac{1}{i}\{i \sinh x \cos y - \cosh x \sin y\}$$

$$[\text{Since } \sin i\theta = i \sinh \theta, \cos i\theta = i \cosh \theta]$$

$$= \sinh x \cos y - \frac{1}{i}\cosh x \sin y$$

$$= \sinh x \cos y + i \cosh x \sin y \; [\text{since } 1/i = -1]$$

Equating the real and imaginary parts, we get

$$u = \sinh x \cos y; \qquad v = \sin y \cos x$$

$$\frac{du}{dx} = \cosh x \cos y; \qquad \frac{dv}{dx} = \sin y \sinh x$$

$$\frac{du}{dy} = -\sin y \, x \sin x; \qquad \frac{dv}{dy} = \cos y \cosh x$$

$$\Rightarrow \frac{du}{dx} = \frac{dv}{dy}; \qquad \frac{du}{dy} = -\frac{dv}{dy}$$

⇒ C-R equations are satisfied.

⇒ f(z) = sinh z is analytic.

3. Let us determine the constants a, b and c if f(z) = x + ay + i(bx + cy) is analytic.

Solution:

Given:

$$f(z) = x + ay + i(bx + cy)$$

Let,

$$f(z) = x + ay + i(bx + cy)$$

$$\Rightarrow u + iv = (x + ay) + i(bx + cy)$$

Equating the real and imaginary parts,

We get,

$$u = x + ay; \qquad v = bx + cy;$$

$$\Rightarrow \frac{du}{dx} = 1; \qquad \frac{dv}{dx} = b$$

$$\frac{du}{dy} = a; \qquad \frac{dv}{dy} = c$$

Since f(z) is analytic, we have,

$$\frac{du}{dx} = \frac{dv}{dy}; \; \frac{du}{dy} = -\frac{dv}{dx}$$

$$\Rightarrow c = 1 \text{ and } a = -b$$

$\Rightarrow$ c = −1 and a = −b , b may be any value.

4. Let us show that f(z) = z^n is differentiable and hence find its derivative.

Solution:

Given:

Let,

$$f(z) = z^n$$

$$P + iQ = \left[re^{i\theta}\right]^n$$

$$= r^n e^{in\theta} = r^n \ [\cos\ n\theta + i\sin\ n\theta]$$

$$= r^n\ \cos\ n\theta + ir^n\ \sin n\theta$$

Equating the real and imaginary parts, we get,

$$p = r^n \cos n\theta; \qquad Q = r^n \sin n\theta$$

$$\frac{\partial p}{\partial r} = nr^{n-1} \cos n\theta; \qquad \frac{\partial Q}{\partial r} = nr^{n-1} \sin n\theta$$

$$\frac{\partial p}{\partial \theta} = r^n(-n \sin n\theta); \qquad \frac{\partial Q}{\partial \theta} = r^n \cdot n\cos n\theta$$

$$\Rightarrow \frac{\partial p}{\partial r} = nr^{n-1} \cos n\theta = \frac{nr^n \cos n\theta}{r} = \left(\frac{\partial Q}{\partial r}\right)\frac{1}{r}$$

$$\frac{\partial Q}{\partial r} = nr^{n-1} \sin n\theta = \frac{nr^n \sin n\theta}{r} = \left(-\frac{\partial Q}{\partial r}\right)\frac{1}{r}$$

$\Rightarrow$ C – R equation are satisfied

$\Rightarrow$ f(z) = z^n is differentiable.

Now,

$$f'(z) = e^{i\theta}\left[\frac{\partial p}{\partial r} + i\frac{\partial Q}{\partial r}\right]$$

$$= e^{-i\theta}\left[nr^{n-1} \cos n\theta + inr^{n-1} \sin n\theta\right]$$

$$= e^{-i\theta}\left(= e^{i\theta}\right)(\cos n\theta + i\sin n\theta)$$

$$= e^{-i\theta}\, nr^{n-1} \cdot e^{n-1}$$

$$= e^{-i} nr^{in\theta} \cdot nr^{n-1} \quad \left[\text{Since } a^m \cdot a^n = a^{m+n}\right]$$

$$= nr^{n-1} e^{i\theta(n-1)}$$

5. Let $z = x + iy$ and $\overline{z} = x - iy$

Solution:

Given:

$$\text{i.e., } x = \frac{z + \overline{z}}{2} \text{ and } y = \frac{z - \overline{z}}{2i}$$

Given,

$W = u + iv$; u and v be functions of x and y.

Consider ω may functions of z $\overline{z}$.

$$\frac{\partial w}{\partial \overline{z}} = \frac{\partial u}{\partial \overline{z}} + i\frac{\partial v}{\partial \overline{z}}$$

$$= \left(\frac{\partial u}{\partial x} \cdot \frac{\partial x}{\partial \overline{z}} + \frac{\partial u}{\partial y} \cdot \frac{\partial y}{\partial \overline{z}}\right) + i\left(\frac{\partial v}{\partial x} \cdot \frac{\partial x}{\partial \overline{z}} + \frac{\partial v}{\partial y} \cdot \frac{\partial y}{\partial \overline{z}}\right)$$

$$= \left(\frac{1}{2}u_x - \frac{1}{2i}u_y\right) + i\left(\frac{1}{2}v_x - \frac{1}{2i}v_y\right)$$

$$= \frac{1}{2}(u_x - v_y) + \frac{i}{2}(u_y + v_x)$$

= 0 (W is analysis if satisfying C.R. equation $u_x - v_y$ and $u_{y=} - v_x$).

Hence W is independent $\overline{z}$ i.e., W is a function of z alone.

3.2 Laplace Equation

Laplace equation can be written as,

$$\nabla^2 u = 0 \text{ in} \subset R^n \qquad ...(1)$$

Where, $\nabla^2 u := \sum_{i=1}^{n} \frac{\partial^2 u}{\partial x_i^2}$ is the Laplacian of the function u. The theory of the solution of the Laplace equation is called as the potential theory. Equation (1) is known as the potential equation, as the function u is frequently a potential function. Solutions of equation (1) that have continuous second order partial derivatives are called as the harmonic functions.

3.2.1 Conformal Mapping

A mapping or transformation which preserves angles in magnitude and in direction between every pair of curves through a point is said to be conformal at that point.

The transformation $w = f(z)$ transforms the curves C_1 and C_2 to the curves C_1^1 and C_2^1 and intersects at a given point then the transformation is said to be a conformal transformation.

The linear function $f(z) = z + b$,

where b is a complex constant, always describes a translation.

- The linear function f(z) = az, a ≠ 0, 1,

where a is a complex constant, always describes a rotation and a magnification.

- Thus the linear function w = f(z) = az + b can be considered as a mapping which comprises of translation, rotation and magnification.

The transformation w = 1/z,

$$f(z) = 1/z \text{ is analytic with } f^1(z) = \frac{-1}{z^2}, \text{ for } z \neq 0.$$

As such, this transformation is conformal at every point. The transformation is carried by taking $z = re^{i\theta}$ and $w = Re^{i\theta}$ in polar form. Then from the above Equation so that a point (r,θ) is transformed to the point $(R,\phi) = \left(1/r, -\theta\right)$.

Example: show that the transformation.

w = 1/z transforms a circle to a circle or a straight line.

Using,

$$W = u + iv, \quad z = x + iy$$

And,

$$w = f(z)$$

$$x + iy = \frac{1}{u + iv} = \frac{u - iv}{u^2 + v^2}$$

Equating real and imaginary parts gives. $x = \frac{u}{u^2 + v^2}, y = \frac{-v}{u^2 + v^2},$ ...(1)

Let us consider any circle in z-plane. Its Cartesian equation is of the form,

$$x^2 + y^2 + 2gx + 2fy + c = 0$$

Substituting for x and y from equation (I), we get,

$$c(u^2 + v^2) + 2gu - 2fu + 1 = 0 \quad ...(2)$$

The above equation represents a circle in the w- plane if and a straight line if c = 0.

Bilinear Transformation or Mobius Transformation

The transformation $w = \frac{az+b}{cz+d'}$ $ad - bc \neq 0$ where a, b, c, d are complex numbers is called a bilinear transformation.

This transformation was first introduced by A.F. Mobius. So it is also called as Mobius transformation.

Note:

- Under bilinear transformation, no two points in z-plane go to the same point in w-plane.
- Bilinear transformation has at most two fixed point or invariant point.
- The bilinear transformation which transforms z_1, z_2, z_3 into w_1, w_2, w_3 is,

$$\frac{(w - w_1)(w_2 - w_3)}{(w - w_3)(w_2 - w_1)} = \frac{(z - z_1)(z_2 - z_3)}{(z - z_3)(z_2 - z_1)}$$

- Given four points z_1, z_2, z_3, z_4 in this order, the ratio $\frac{(z_1 - z_2)(z_3 - z_4)}{(z_2 - z_3)(z_4 - z_1)}$ is called the cross ratio of the points.
- It can be expressed as cwz + dw – (az + b) = 0. It is linear in both w and z.

∴ It is called as bilinear transformation.

- Bilinear transformation is conformal only when $\frac{dw}{dz} \neq 0$.

$$\frac{ad - bc}{(cz + d)^2} \neq 0$$

$ad - bc \neq 0$

$ad - bc = 0$ very point in the z-plane is a critical point

- The inverse of the bilinear transformation $w = \frac{az+b}{cz+d}$ is $z = \frac{-dw+b}{cw-a}$ which is also a bilinear transformation except $w = \frac{a}{c}$.
- Each point in the plane except $z = \frac{-d}{c}$ corresponds to a unique point in the w-plane.

The point $z = \frac{-d}{c}$ corresponds to a infinity in the w-plane.

- The cross ratio of four points $\frac{(w_1 - w_2)(w_3 - w_4)}{(w_2 - w_3)(w_4 - w_1)} = \frac{(z_1 - z_2)(z_3 - z_4)}{(z_2 - z_3)(z_4 - z_1)}$ is invariant under bilinear transformation.
- If one of the point is the point at infinity the quotient of those difference which involve this point is replaced by 1.

Problems based on Bilinear Transformation

1. Let us determine the fixed point of $W = \frac{2zi+5}{z-4i}$

Solution:

$$z = \frac{2zi+5}{z-4i}$$

$$z^2 - 4iz = 2zi + 5$$

$$z^2 - 6iz - 5 = 0$$

$$z = \frac{6i \pm \sqrt{-36+20}}{2}$$

$$z = 5i, i$$

2. We shall determine the bi-linear transformation which maps the points z = 1, i, - 1 into the points ω = i, 0, - i. Hence find the image of | z | < 1.

Solution:

Bi-linear transformation is,

$$\frac{(w-w_1)(w_2-w_3)}{(w-w_3)(w_2-w_1)}=\frac{(z-z_1)(z_2-z_3)}{(z-z_3)(z_2-z_1)}.$$

$$\Rightarrow\frac{(w-i)(o+i)}{(w-i)(o-i)}=\frac{(z-1)(i+1)}{(z+1)(i-1)}.$$

$$\frac{(w-i)(\not{i})}{(w+i)(-\not{i})}=\frac{(z-1)(i+1)}{(z+1)(i-1)}$$

$$\frac{(w-i)}{(w+i)}=\frac{(iz-1)}{(z+1)}$$

$$\frac{(w-i)+(w+i)}{(w-1)-(w+i)}=\frac{(iz-i)+(z+1)}{(iz-i)-(z+1)}$$

$$\Rightarrow\frac{2w}{-2i}=\frac{z(i+1)+(1\text{-}i)}{z(i-1)+(-1-i)} \qquad (1-i)=-i(i+1)$$

$$\frac{w}{-i}=\frac{z(i+1)-i(i+1)}{zi(i+1)-(i+1)} \qquad i-1=i(i+1)$$

$$w_{-1}=\frac{z-i}{zi-1}\Rightarrow w=-\frac{-iz-1}{zi-1}$$

$$w=\frac{iz+1}{1-zi}\Rightarrow w=\frac{i(z+1/i)}{i(-z+1/i)}=\frac{i(z-i)}{i(-1-z)}$$

$$w=\frac{z-i}{-i-z}\Rightarrow w=\frac{-z+i}{z+i}$$

To find image of | z | < 1,

$$w=\frac{-z+i}{z+i}\Rightarrow wz+wi=-z+i$$

$$wz+z+i=-wi\Rightarrow z=\frac{i-wi}{w+1}=i\left(\frac{1-w}{w+1}\right)$$

$$z = i\left(\frac{-w+1}{w+1}\right)$$

Given,

$$|z|<1 \Rightarrow \left|\frac{i(-w+1)}{w+1}\right|<1$$

$$\Rightarrow |-wi+i| < |w+1|$$

$$\Rightarrow |-(u+iv)i+i| < |u+iv+1|$$

$$\Rightarrow |-ui+v+i| < |(u+1)+iv|$$

$$\Rightarrow |v+i(1-u)|^2 < |(u+1)+iv|^2$$

3. Here we shall determine the bi-linear transformation which maps the points z = 0, 1, ∞ into ω = i, 1, – i respectively.

Solution:

Given:

$$z_1 = 0 \;;\; z_2 = 1 \;;\; z_3 = \infty$$

$$\omega_1 - i \;;\; \omega_2 = 1 \;;\; \omega_3 = -i.$$

Here,

$$z_3 = \infty \Rightarrow \frac{1}{z_3} \to 0. \text{ (1) Becomes,}$$

$$\frac{(w-w_1)}{(w-w_3)}\frac{(w_2-w_3)}{(w_2-w_1)} = \frac{(z-z_1)}{z_3\left(\frac{z}{z_3}-1\right)}\frac{z_3\left(\frac{z_2}{z_3}-1\right)}{(z_2-z_1)}$$

$$\frac{(w-i)}{(w+i)}\frac{(1+i)}{(1-i)} = \frac{(z-0)}{(0-1)}\frac{(0-1)}{(1-0)}$$

$$\frac{(w-i)}{(w+i)}(i) = z\left(\because \frac{1+i}{1-i} = i\right)$$

$$i\omega + 1 = \omega z + iz$$

$$i\omega - \omega z = iz - 1$$

$$\omega(i - z) = iz - 1$$

i.e., $w = \dfrac{iz-1}{-z+i}$ which is the required bi-linear transformation,

$$\Rightarrow v^2 + (1-u)^2 < (u+1)^2 + v^2$$

$$1 + u^2 - 2u < u^2 + 1 + 2u.$$

$$4u > 0 \Rightarrow u > 0$$

Is the required image of $|z| < 1$.

4. We shall calculate the bi-linear transformation which maps the points $z = 0, 1, \infty$ into $\omega = i,\ 1, -i$ respectively.

Solution:

Given:

$$z_1 = 0\ z_2 = 1\ z_3 = \infty$$

$$\omega_1 = i\ \omega_2 = 1\ \omega_3 = -i$$

Bi-linear transformation is,

$$\frac{(w-w_1)(w_2-w_3)}{(w-w_3)(w_2-w_1)} = \frac{(z-z_1)(z_2-z_3)}{(z-z_3)(z_2-z_1)}$$

$$\Rightarrow \frac{(w-i)(1+i)}{(w+i)(1-i)} = \frac{z-0}{1-0}\left(\because z_3 = \infty\right)$$

$$\Rightarrow \frac{(w-i)\,i(1-i)}{(w+i)\,(1-i)} = z$$

$$(w-i)i = z(\omega+i)$$

$$\omega i + 1 = z\omega + zi$$

$$\omega i - z\omega = zi - 1$$

$$\omega(i-z)=zi-1$$

$$\omega=\frac{zi-1}{1-z}=\frac{i(z+i)}{i(1+zi)}$$

$$\omega=\frac{z+i}{1+zi}.$$

5. Find the bi-linear transformation that transforms 1, i and - 1 of the z-plane onto 0, 1 and of the ω-plane. Also show that the transformation maps interior of the unit circle of the z-plane on to upper half of the co-plane.

Solution:

Given:

$$z_1=1;\ z_2=i;\ z_3=-1$$

$$w_1=0;\ w_2=1;\ \&\ w_3=\infty$$

$$\frac{(w-w_1)}{(w-w_3)}\frac{(w_2-w_3)}{(w_2-w_1)}=\frac{(z-z_1)}{(z-z_3)}\frac{(z_2-z_3)}{(z_2-z_1)} \qquad ...(1)$$

Here,

$$w_3=\infty\Rightarrow\frac{1}{w_3}\to 0$$

Rewrite (1),

$$\frac{(w-w_1)}{w_3\left(\frac{w}{w_3}-1\right)}\frac{w_3\left(\frac{w_2}{w_3}-1\right)}{(w_2-w_1)}=\frac{(z-z_1)}{(z-z_3)}\frac{(z_2-z_3)}{(z_2-z_1)}$$

$$\frac{(w-0)}{(0-1)}\frac{(0-1)}{(1-0)}=\frac{(z-1)(i+1)}{(z+1)(i-1)}$$

$$w=\frac{(z-1)}{(z+1)}(-i)$$

$$=\frac{-iz+i}{z+1} \qquad ...(2)$$

Which is the required bi-linear transformation, inverse transformation,

$$\omega z + \omega + iz = i$$

$$z(\omega + i) = (i - \omega)$$

$$z = \frac{i - w}{w + 1} \qquad \dots(3)$$

The image of interior of the unit circle,

i.e., $|z| < 1$ is,

$$\left|\frac{i-\omega}{\omega+i}\right| < 1$$

$$|i - \omega| < |\omega + i|$$

$$|i - (u + iv)| < |u + iv + i|$$

$$|-u + i(1 - v)| < |u + i(v + 1)|$$

$$u^2 + (1 - v)^2 < u^2 + (v + 1)^2$$

$$u^2 + 1 - 2v + v^2 < u^2 + v^2 + 2v + 1$$

i.e., $4v > 0$

$v > 0$ Which is upper half of the ω-plane.

Conformal Mapping: A mapping or transformation which preserves angles in magnitude and in direction between every pair of curves through a point is said to be conformal at that point.

Isogonal Transformation: The transformation preserves the angle in magnitude but not in direction between every pair of curves through a point is said to be a isogonal at that point.

Critical Point: A point at which f' (z) = 0 is called a critical point of the transformation.

At the critical point, the transformation w = f (z) is not conformal.

1. Let us determine the critical points for the transformation $w^2 = (z - \alpha)(z - \beta)$.

Solution:

$$w^2 = (z - \alpha)(z - \beta)$$

$$2w\frac{dw}{dz}=(z-\alpha)+(z-\beta)=2z-(\alpha+\beta)$$

$$\therefore w\frac{dw}{dz}=z-\frac{1}{2}(\alpha+\beta)$$

Critical points occur at $\frac{dw}{dz}=0,$

$$\therefore z--(\alpha+\beta)=0$$

Also,

$$\frac{dw}{dz}=\frac{w}{z-\frac{1}{2}(\alpha+\beta)}$$

The Critical points occurs at $\frac{dw}{dz}=0,$

$$\therefore \frac{w}{z-\frac{1}{2}(\alpha+\beta)}=0$$

$$\Rightarrow w=0$$

$$\Rightarrow (z-\alpha)(z-\beta)=0$$

$$\Rightarrow z=\alpha, z=\beta$$

$\therefore$ The Critical points occurs at $\therefore z=\frac{1}{2}(\alpha+\beta), \alpha \,\&\, \beta$

2. Let us determine the image of the circle | z |= c by the transformation w = 5z.

Solution:

Given:

$$w = 5z$$

$$u+iv=5(x+iy)$$

$$=5x+i5y$$

$$\Rightarrow u=5x, v=5y3$$

Given:

$$|z|=C$$

$$\Rightarrow \left(x^2 + y^2\right)^{/} = C$$

$$\Rightarrow x^2 + y^2 = C^2$$

$$\Rightarrow \left(\frac{u}{5}\right)^2 + \left(\frac{v}{5}\right)^2 = C^2$$

$$\Rightarrow u^2 + v^2 = 25C^2$$

$$\Rightarrow u^2 + v^2 = (5C)^2$$

z = C Maps to a circle in w-plane with centre at the origin and radius 3C.

3. Here we shall show that the image of the hyperbola $x^2 - y^2 = 1$ under the transformation $w = \frac{1}{z}$ is the lemniscate $r^2 = \cos 2\theta$.

Solution:

Given:

$$w = \frac{1}{z} \qquad w = u + iv \qquad z = x + iy.$$

$$z = \frac{1}{w} = \frac{1}{u+iv}\left(\frac{u-iv}{u-iv}\right) = \frac{u-iv}{u^2+v^2}$$

$$x + iy = \frac{u-iv}{u^2+v^2} \Rightarrow x = \frac{u}{u+v^2} \;\&\; y = \frac{-v}{u^2+v^2}.$$

Now,

$$x^2 - y^2 = 1 \qquad \ldots(1) \text{ Becomes,}$$

$$\frac{u^2}{\left(u^2+v^2\right)^2} - \frac{v^2}{\left(u^2+v^2\right)^2} = 1$$

$\therefore\ u^2 - v^2 = \left(u^2+v^2\right)^2$ is the image of (1) under $\omega = 1/z$.

Let us convert the image in to polar form.

Let,

$$u = r\ \cos\ \theta\ v = r\ \sin\ \theta.$$

Then,

$$r^2 \cos^2 \theta - r^2 \sin^2 \theta = \left(r^2 \cos^2 \theta + r^2 \sin^2 \theta\right)^2$$

$$r^2 \left(\cos^2 \theta - \sin^2 \theta\right) = \left[r^2 (\cos^2 \theta + \sin^2 \theta)\right]^2$$

$$\left[r^2 \left(\cos^2 \theta + \sin^2 \theta\right)\right]^2$$

$\therefore\ r^2 = \cos 2\theta$ is the image of (1) in polar form.

4. Let us determine the image of | z - 2i | = 2 under the transformation.

Solution:

The image of $|Z - 2i| = 2$ under $w = \frac{1}{z}$ is $|\frac{1}{w} - 2i| = 2$

i.e.,

$$|1 - 2iW| = 2|W|$$

$$|1 - 2i(u + iv)| = 2|u + iv|$$

$$|(1 + 2v) - 2iu| = 2|u + iv|$$

$$\sqrt{(1+2v)^2 + 4u^2} = 2\sqrt{u^2 + v^2}$$

Squaring both sides,

$$(1 + 2v)^2 + 4u^2 = 4\left(u^2 + v^2\right)$$

i.e.,

$$1 + 4v + 4v^2 + 4u^2 = 4u^2 + 4v^2$$

$$1 + 4v = 0$$

$$\Rightarrow v = -\frac{1}{4}.$$

i.e., the image of $|Z - 2i| = 2$ under $w = \frac{1}{z}$ is a straight line $\Rightarrow v = -\frac{1}{4}$.

5. Let us calculate the image of the half-plane x c, c o under $w=\frac{1}{z}$ Sketch graphically. Also determine the fixed points of .

Solution:

Given:

$$w=\frac{1}{z}$$

$$u+iv=\frac{1}{x+iy}$$

$$=\frac{(x-iy)}{(x+iy)(x-iy)}=\frac{x}{x^2+y^2}-i\frac{y}{x^2+y^2}$$

$$u=\frac{x}{x^2+y^2} \qquad ...(1)$$

$$v=\frac{-y}{x^2+y^2} \qquad ...(2)$$

Now,

$$w=\frac{1}{z}\Rightarrow z=\frac{1}{w} \qquad ...(3)$$

$$\therefore x=\frac{u}{u^2+v^2};y=\frac{-v}{u^2+v^2} \qquad ...(4)$$

Now the image of the half plane,

$$x>C \text{ is } \frac{u}{u^2+v^2}>C$$

(i.e.) $u^2+v^2<\frac{u}{c}$; $u^2-\frac{u}{c}+v^2<0$

$$\left(u-\frac{1}{2c}\right)^2+v^2<0\left(\frac{1}{2c}\right)^2$$

Which is interior of the circle whose center is $\left(\frac{1}{2c}, 0\right)$ and radius $\frac{1}{2c}$.

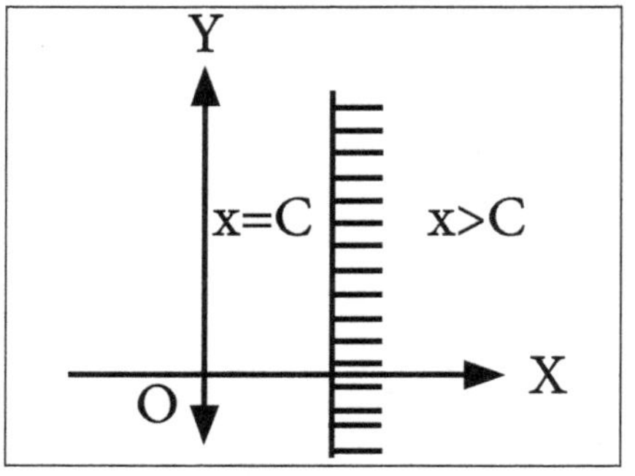

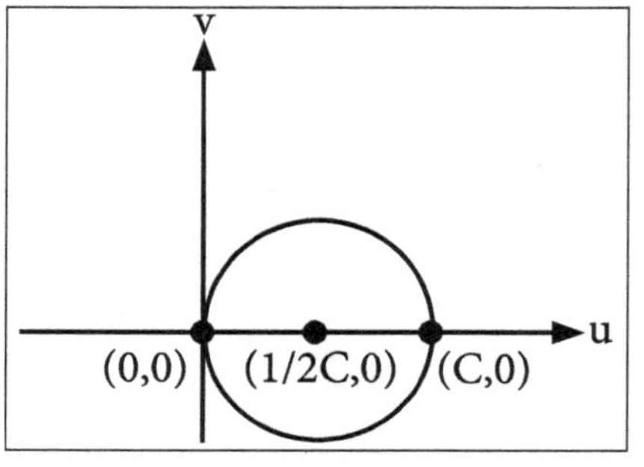

Fixed point of $\omega = f(z)$ is $z = f(z)$ (image of a point is itself).

$$\frac{1}{z} = z \Rightarrow z^2 = 1$$

i.e., $z = \pm 1$ which is fixed point of $w = \frac{1}{z}$.

3.3 Complex Integration: Line Integral in the Complex Plane

For some special functions and domains the integration is path independent but this should not be taken to be the case in general. Given the sensitivity of the path taken for a given integral and its result parametrization is often the most convenient way to evaluate such integrals. Complex variable techniques have been used in a wide variety of areas of engineering. This has been particularly true in areas such as electromagnetic field theory, aerodynamics, fluid dynamics and elasticity.

With the rapid developments in computer technology and the consequential use of sophisticated algorithms for analysis and design in engineering there has been in recent years less emphasis on the use of complex variable techniques and a shift towards numerical techniques applied directly to the underlying full partial differential equations model of the situation. However it is useful to have an analytic solution possibly for an idealized model in order to develop a better understanding of the solution and to develop confidence in numerical estimates for the solution of more sophisticated models.

The design of aerofoil sections for aircraft is an area where the theory was developed using complex variable techniques. Throughout engineering, transforms defined as complex integrals in one form or another play a major role in analysis and design. The use of complex variable techniques allows us to develop criteria for the stability of systems.

Line Integral in the Complex Plane

Complex definite integrals are called as the line integrals and are written as,

$$\int_C f(z)dz$$

The integrand f(z) is integrated over a given curve C in the complex plane called the path of integration, normally represented by a parametric representation,

$$z(t)=x(t)+iy(t), \qquad a\le t\le b$$

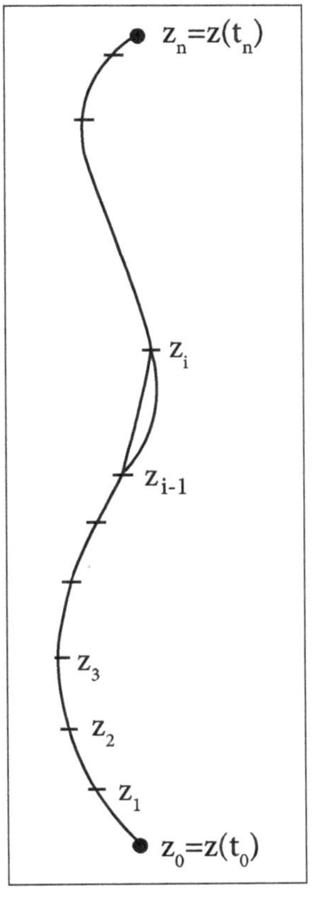

The sense of increasing t is called as the positive sense on C. The curve C is assumed to be a smooth curve, (i.e.) it has continuous and non-zero derivative at each $t \in (a, b)$. In case, if the initial point and terminal point of a curve coincide, (i.e.) $z(a) = z(b)$, the curve is said to be a closed one.

The Complex Line Integral

Consider a smooth curve C in the complex plane given by, $z(t) = x(t) + iy(t)$, $a \le t \le b$ and f(z) a continuous function defined at each point of C. Divide C into n parts at the points $z_o, z_1, \ldots, z_{n-1}, z_n$ corresponding to the partition $t_o(= a), t_1, \ldots, t_{n-1}, t_{n\,(= b)}$, such that, $t_o < t_1 < \ldots < t_n$ of the interval $a \le t \le b$, as shown in the above figure.

Let,

$$\Delta z_i = z_i - z_{i-1} \text{ and } \xi_i \text{ be any point on the arc } z_{i-1}\,z_i \text{ between } z_{i-1} \text{ and } z_i.$$

Consider the sum,

$$S_n = \sum_{i=1}^{n} f(\xi_i)\Delta z_i$$

The limit of the sum S_n as $n \to \infty$ in such a way that the length of the chord Δz_i approaches zero is called as the line integral of f(z), taken along the curve C oriented from z_o to z_n and is denoted by,

$$I = \int_C f(z)dz \qquad \ldots(1)$$

In case the path of integral C is a closed curve, then the integration is denoted by $\oint_C f(z)dz$ and the integral along a closed curve is sometimes called as the contour integral.

In general, the path of integration for complex line integrals are assumed to be piecewise smooth, (i.e.), consisting of finitely smooth curves joined end to end.

Writing $f(z) = u(x, y) + iv(x, y)$ and $dz = dx + idy$, the line integral of equation (1) can be expressed as,

$$\int_C f(z)dz = \int_C (u+iv)(dx+idy)$$

$$= \int_C (udx - vdy) + i\int_C (vdx + udy) \qquad \ldots(2)$$

Thus, the line integral of a complex function can be evaluated in terms of two line integrals of real functions.

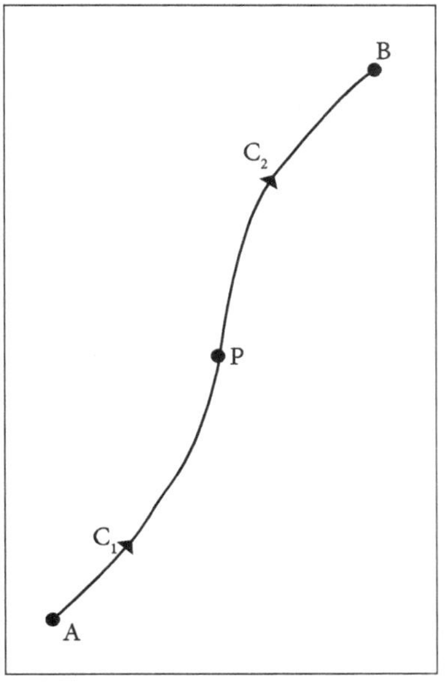

Basic Properties of Line Integrals

A few basic properties of the line integrals of a complex function are given below:

Linearity:

$$\int_C [k_1 f_1(z) + k_2 f_2(z)] dz = k_1 \int_C f_1(z) dz + k_2 \int_C f_2(z) dz$$

Where, k_1 and k_2 are two constants.

Sense reversal:

$$\int_{z_0}^{z} f(z) dz = -\int_{z}^{z_0} f(z) dz$$

Partitioning of path:

$$\int_C f(z) dz = \int_{C_1} f(z) dz + \int_{C_2} f(z) dz$$

Where, the curve C consists of two smooth curves C_1 and C_2 joined end to end as shown in the above figure.

ML inequality:

$$\left| \int_C f(z) dz \right| \leq ML$$

Where,

M = Constant, such that $|f(z)| \leq M$ everywhere on C.

L = Length of the curve.

3.3.1 Cauchy's Integral Theorem and Cauchy's Integral Formula

Statement

If f(z) is analytic and f'(z) is continuous at all points inside and on a simple closed curve c, then

$$\int_C f(z)dz = 0$$

Proof: Let R be the region by c,

$$z = x + iy,\ dz = dx + idy,\ f(z) = u + iv$$

Now,

$$= \int (udx - vdy) + i\int (vdx + udy) \quad ...(1)$$

Since f'(z) is continuous the four partial derivatives $\frac{\partial u}{\partial x}, \frac{\partial u}{\partial y}, \frac{\partial v}{\partial x}, \frac{\partial v}{\partial y}$ are also exists and continuous in R and on c.

By Green's theorem in the plane,

$$\int_C Mdx + Ndy = \int\int_R \left(\frac{\partial N}{\partial x} - \frac{\partial M}{\partial y}\right)dxdy$$

∴ (1) Becomes,

$$\int_C f(z)dz = \iint_R \left(-\frac{\partial v}{\partial x} - \frac{\partial u}{\partial y}\right)dxdy + i\iint_R \left(-\frac{\partial v}{\partial x} - \frac{\partial u}{\partial y}\right)dxdy \quad ...(2)$$

Since f(z) = u + iv is analytic by the cauchy's equation,

∴ From (2)

$$\int_C f(z)dz = \iint_R \left(\frac{\partial v}{\partial x} - \frac{\partial u}{\partial y}\right)dxdy + i\iint_R \left(-\frac{\partial v}{\partial x} - \frac{\partial u}{\partial y}\right)dxdy$$

$$= 0 + i0 = 0$$

Cauchy's Integral Formula or Cauchy's Fundamental Formula

Statement:

If f(z) is analytic inside and on a simple closed curve c and if 'a' is any point within c then,

$f(a)=\frac{1}{2\pi i}\int_c \frac{f(z)}{z-a}dz$, the integration around c being taken in the positive direction.

Proof:

Given f(z) is analytic inside and on c. Now $\frac{f(z)}{z-a}$ is analytic inside and on c except at z = a. Draw a circle with center at z = a and radius r units such that c_1 lies entirely inside c.

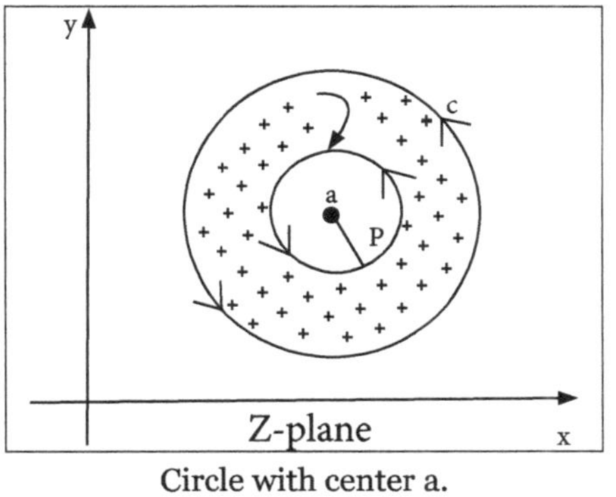

Circle with center a.

Now $\phi(z)=\frac{f(z)}{z-a}$ is analytic in the region enclosed between c and c_1,

$$\therefore \int_c \phi(z)dz=\int_{c_1}\phi(z)dz$$

i.e., $$\therefore \int_c \frac{f(z)}{z-a}dz=\int_{c_1}\frac{f(z)}{z-a}dz$$

On c_1, any point z is given by $z=a+re^{i\theta}, 0\le\theta\le 2\pi \quad \therefore dz=ire^{i\theta}d\theta$

$$\therefore \int_c \frac{f(z)}{z-a}dz=\int_0^{2\pi}\frac{f\left(a+re^{i\theta}\right)}{re^{i\theta}}ire^{i\theta}d\theta$$

$$=2\pi i f(a)$$

$$\therefore f(a)=\frac{1}{2\pi i}\int_C \frac{f(z)}{z-a}dz$$

Problems

1. Using Cauchy's integral formula let us evaluate $\int \frac{z}{z^2+1} dz$, where C is the circle $|z+i|=1$.

Solution:

Cauchy's integral formula is,

$$\int_C \frac{f(Z)}{z-z_0} dz = 2\pi i f(z_0)$$

$$\frac{1}{z^2+1} = \frac{1}{z^2+i^2} = \frac{1}{(z+i)(z-i)} = \frac{A}{z+i} + \frac{B}{z-i}.$$

$$1 = A(z-i) + B(z+i).$$

$$z = -i \Rightarrow 1 = -2iA \Rightarrow A = \frac{-1}{2i} = \frac{i}{2}.$$

$$z = +i1 = 2iB \Rightarrow B = \frac{1}{2i} = -\frac{i}{2}$$

$$\therefore \frac{1}{z^2+1} = \frac{i/2}{z+i} + \frac{-i/2}{z-i} = \frac{i}{2}\left(\frac{1}{z+i} - \frac{1}{z-i}\right)$$

$$\therefore I = \int_C \frac{z}{z^2+1} dz = \int_C z\left(\frac{i}{2}\left(\frac{1}{z+i} - \frac{1}{z-i}\right)\right) dz.$$

$$= \frac{i}{2}\int_C \frac{z}{z+i} dz - \frac{i}{2}\int_C \frac{z}{z-i} dz = \frac{i}{2} I_1 - \frac{i}{2} I_2$$

C is $|z+i|=1.$

Here,

$z=-i$ lies inside C and $z=i$ lies outside C.

$$I_1 = \int_C \frac{z}{z+i} dz = 2\pi i f(-i) = 2\pi i(-i) = 2\pi$$

$$I_2 = \int_C \frac{z}{z-i} dz = 0$$

Since $z = i$ lies outside C and hence by Cauchy's theorem.

$$\therefore I = \frac{i}{2}(2\pi) - \frac{i}{2}(0)$$

$$I = \pi i.$$

2. Using Cauchy's integral formula, let us evaluate $\int_C \frac{4-3x}{z(z-1)(z-2)} dx$, where 'C' is the circle $|z| = \frac{3}{2}$.

Solution:

Given:

$$\int_C \frac{4-3x}{z(z-1)(z-2)} dx, \quad |z| = \frac{3}{2}$$

Formula to be used in Cauchy's integral,

$$\int_C \frac{f(x)}{(z-a)} dz = 2\pi \text{ if } (a) \text{ if } z = a \text{ lies inside C.}$$

Consider,

$$\frac{4-3x}{z(z-1)(z-2)} = \frac{A}{z} + \frac{B}{z-1} + \frac{C}{z-2} \qquad ...(1)$$

$$4 - 3z = A(z-1)(z-2) + Bz(z-2) + Cz(z-1) \quad ...(2)$$

Put,

$$z = 0 : 4 = 2A \Rightarrow A = 2.$$

$$z = 1 : 1 = -B \Rightarrow B = -1.$$

$$z = 2 : -2 = 2C \Rightarrow C = -1.$$

$$\frac{4-3x}{z(z-1)(z-2)} = \frac{2}{z} - \frac{1}{z-1} - \frac{1}{z-2}$$

i.e.,

$$\int_C \frac{4-3x}{z(z-1)(z-2)} dz = 2\int_C \frac{1}{z} dz - \int_C \frac{dz}{z-1} - \int_C \frac{dz}{z-2}$$

- z = 0 lies inside C = 3/2.
- z = 1 lies inside C.

- z = 2 lies outside C.

∴ By Cauchy's integral formula,

$$\int_C \frac{4-3x}{z(z-1)(z-2)} dz = 2(2\pi i)f(0) - 2\pi\, if(1) - 0$$

Where,

$$f(z) = 1$$

$$= 4\pi - 2\pi i$$

$$= 2\pi i.$$

3. Let us evaluate $\int_C \frac{zdz}{(z-1)(z-2)^2}$ where C is $|z-2| = \frac{1}{2}$ by using cauchy's integral formula.

Solution:

Let,

$$I = \int_C \frac{zdz}{(z-1)(z-2)^2}.$$

Hence,

z = 1 and z = 2 are singular points of $\frac{z}{(z-1)(z-2)^2}$.

C is $|z-2| = 1/2$.

z = 1 lies outside C and z = 2 lies inside C.

$$\therefore I = \int_C \frac{\left(\frac{z}{z-1}\right)}{(z-2)^2} dz$$

Now,

$f(z) = \frac{oz}{z-1}$ is analytic inside and on C.

WKT $f(a) \quad \frac{}{2 \ \ i}\int\frac{f(z)}{(z \quad a)}dz$ if f(z) is analytic inside and on C.

$$\therefore I = 2\pi \, \text{if}(2), f(z) = \frac{z}{z-1}.$$

$$f(z) = \frac{(z-1)(1)-z(1)}{(z-1)^2} = \frac{z-1-z}{(z-1)^2} = \frac{-1}{(z-1)^2}$$

$$f(z) = \frac{-1}{(2-1)^2} = -1$$

$$\therefore I = 2\pi i(-1)$$

$$= -2\pi i$$

$$\therefore \int_C \frac{zdz}{(z-1)(z-2)^2} = 2\pi i.$$

4. Let us evaluate $\int_C \frac{z+4}{z^2+2z+5}dz$, where c is the circle $|z+1+i| = 2$, using Cauchy's integral formula.

Solution:

Given:

$$\int_C \frac{z+4}{z^2+2z+5}dz,$$

We know that Cauchy's integral formula,

$$\int_C \frac{f(z)}{(z-a)}dz = 2\pi \, \text{if}(a) \text{ if } z = a \text{ lies inside C}$$

Consider the poles are $z^2 + 2z + 5 = 0$,

$$\Rightarrow z = -1 \pm 2i$$

$$z = -1 + 2i = \alpha \text{ say}$$

$$z = -1 - 2i = \beta \text{ say.}$$

Here,

$$z = -1 + 2i(\alpha) : |-1-2i+1+i| = |3i| > 2 \text{ lies outside C.}$$

$z=-1-2i(\beta):|-1-2i+1+i|=|-i|<2$ lies inside C.

$$\therefore \int_C \frac{z+4}{z^2+2z+5}dz = \int_C \frac{\frac{z+4}{(z-\alpha)}}{(z-\beta)}dz \text{ where } f(z)=\left(\frac{z+4}{z-\alpha}\right)$$

$=2\pi \text{ if } (\beta)$ by Cauchy's formula

$$=2\pi i\left(\frac{\beta+4}{\beta-\alpha}\right)$$

$$=2\pi i\frac{(3-2i)}{-4i}=\frac{\pi}{2}(3-2i).$$

5. Let us evaluate $\int_C \frac{z+1}{(z^2+2z+4)^2}dz$, where C is the circle $|z+1+i|=2$, by Cauchy's integral formula.

Solution:

Given:

$$\int_C \frac{z+1}{(z^2+2z+4)^2}dz,$$

Poles of the f(z) are,

$$(z^2+2z+4)^2=0 \Rightarrow (z-a)^2(z-b)^2=0.$$

$z=-1\pm i\sqrt{3}$ twice.

Given,

$$C:|z+1+i|=2.$$

i. The pole

$$z=-1\pm i\sqrt{3}:1-1\pm i\sqrt{3}+1+i1=1(1+\sqrt{3})i1$$

i.e., The pole z = a lies outside c.

ii. The pole z = b lies inside c.

$$\therefore I=\int_C \frac{z+1}{(z-a)^2(z-b)^2}\int_C \frac{\frac{z+1}{(z-a)^2}}{(z-b)^2}dz$$

$$\therefore I = \frac{2\pi i}{1!}\left\{\frac{d}{dz}\frac{(z+1)}{(z-a)^2}\right\}_{z=b} \qquad ...(1)$$

Now,

$$\frac{d}{dz}\frac{(z+1)}{(z-a)^2} = \frac{(z-a)^2(1)-(z+1)2(z-a)}{(z-a)^4} = \frac{(z-a)[z-a-2z-2]}{(z-a)^4}$$

$$= -\frac{(z+a+2)}{(z-a)^3}$$

$$\left[\frac{d}{dx}\frac{z+1}{(z-a)^2}\right]_{z=b} = \frac{-(b+a+2)}{(b-a)^3} = \frac{\left[-1-i\sqrt{3}-1+i\sqrt{3}+2\right]}{\left[-1-i\sqrt{3}+1+i\sqrt{3}\right]^3}$$

$$= 0.$$

$$I = 2\pi i(0) = 0.$$

3.3.2 Derivatives of Analytic Functions

Constructions of Analytic Functions (Milne- Thomson Method):

i. To find f(z) when u given,

$$f'(z) = \frac{\partial u}{\partial x} + i\frac{\partial v}{\partial x}$$

$$= \frac{\partial u}{\partial x} - i\frac{\partial v}{\partial y}$$

$$\phi\ (z,0) = \left(\frac{\partial}{\partial}\right)_{(z,0)}$$

$$\phi_2\ z,0 = \left(\frac{\partial u}{\partial y}\right)_{(z,0)}$$

$$f'(z) = \phi_1(z,0) - i\phi_2(z,0)$$

$$\int f'(z)\,dz = \int \phi_1(z,0)\,dz - i\int \phi_2(z,0)\,dz$$

$$f(z)=\int\phi_1(z,o)dz-i\int\phi_2(z,o)dz+C$$

ii. To find f(z) when v given,

$$f'(z)=\frac{\partial v}{\partial y}+i\frac{\partial v}{\partial x}$$

$$\phi_1(z,o)=\left(\frac{\partial v}{\partial y}\right)_{(z,o)}$$

$$f'(z)=\phi_1(z,o)+i\phi_2(z,o)$$

$$\int f'(z)dz=\int\phi_1(z,o)dz+i\int\phi_2(z,o)dz$$

$$f(z)=\int\phi_1(z,o)dz+i\int\phi_2(z,o)dz+C.$$

Problems

Problems Based on Constructions of Analytic Functions (Milne-Thomson Method).

1. We shall determine an analytic function whose real part is ex(xcosy - ysiny).

Solution:

Given:

$$u=e^x(x\cos y-y\sin y)$$

$$\phi_1(x,y)=\frac{\partial u}{\partial x}=e^x(\cos y)+(x\cos y-y\sin y)e^x$$

$$\phi_1(z,o)=ez+ze^z$$

$$\phi_2(x,y)=\frac{\partial u}{\partial y}=e^x(-x\sin y)-(y\cos y+\sin y)e^x$$

$$\phi_2(z,o)=o$$

$$f(z)=\int\phi_1(z,o)dz-i\int\phi_2(z,o)dz+c$$

$$=\int e^z+ze^z dz-i\int o\,dz+c$$

$$= \int (1+z)e_z \, dz$$

2. Let us determine an analytic function whose real part is $u = x^3 - 3xy^2 + 3x^2 - 3y^2 + 1$.

Solution:

Given:

$$u = x^3 - 3xy^2 + 3x^2 - 3y^2 + 1$$

$$\phi_1(x,y) = \frac{\partial u}{\partial x} = 3x^2 - 3y^2 + 6x$$

$$\phi_1(z,o) = 3z^2 + 6z$$

$$\phi_2(x,y) = \frac{\partial u}{\partial x} = 3x^2 - 3y^2 + 6x$$

$$\phi_2(z,\ o) = o$$

$$f(z) = \int \phi_1(z,\ o)dz - i\int \phi_2(z,\ o)dz + c$$

$$= \int 3z^2 + 6z - i\int odz + c$$

$$= z^3 + 3z^2 + c$$

3. Here we shall determine an analytic function $f(z) = u + iv$ where $u + v = \dfrac{x}{x^2 + y^2}$.

Solution:

Given:

$$f(z) = u + iv$$

$$u + v = \frac{x}{x^2 + y^2}$$

$$if\ (z) = iu - v$$

$$(u - v) + i(u + v) = (i + 1)f(z)$$

$$U = (u - v)$$

$$V = (u - v)$$

$$F(z)=(i+z)f(z)$$

Given,

$$V=\frac{x}{x^2+y^2}$$

$$\phi_1(x,y)=\left(\frac{\partial v}{\partial y}\right)=\frac{(x^2+y^2)0-x(2y)}{(x^2+y^2)^2}=\frac{-x(2y)}{(x^2+y^2)^2}$$

$$\phi_1(z,0)=\left(\frac{\partial v}{\partial y}\right)_{(z,0)}=0$$

$$\phi_2(x,y)=\left(\frac{\partial v}{\partial x}\right)=\frac{(x^2+y^2)1-x(2x)}{(x^2+y^2)^2}=\frac{y^2-x^2}{(x^2+y^2)^2}$$

$$\phi_2(x,y)=\left(\frac{\partial v}{\partial x}\right)_{(z,0)}=\frac{-1}{z^2}$$

$$F'(z)=\phi_1(z,0)+i\phi_2(z,0)$$

$$\int F'(z)dz=\int\phi_1(z,0)dz+i\int\phi_2(z,0)dz$$

$$=0+i\int\frac{-1}{z^2}dz$$

$$=i\frac{1}{z^2}+c$$

$$(1+i)f(z)=i\frac{1}{z^2}+c$$

$$f(z)=\frac{1+i}{2z}+c.$$

4

Power Series

4.1 Power Series, Taylor's Series and Laurent's Series

A Taylor series is a representation of a function as an infinite sum of terms that are calculated from the values of the function's derivatives at a single point.

Taylor's Series Method

Consider the first order differential equation,

$$\frac{dy}{dx}=f(x,y),y(x_o)=y_o$$

The solution of the above initial value problem is obtained in two types as,

- Power series solution.
- Point wise solution.

Power Series Solution

$$y(x)=y_o+\frac{x-x_o}{1!}y_o'+\frac{(x-x_o)^2}{2!}y_o''+\frac{(x-x_o)^3}{3!}y_o'''+....$$

Where,

$$y_m^{(r)}=\frac{d^ry}{dx^r}\text{ at }(x_m,y_m)$$

Point Wise Solution

$$y_{m+1}=y_m+\frac{h}{1!}y_m'+\frac{h^2}{2!}y_m''+\frac{h^3}{2!}y_m'''+...$$

Where,

$$y_m^{(r)}=\frac{d^ry}{dx^r}\text{ at }(x_m,y_m)$$

And $m=0,1,2.......$

Taylor's Series

Statement: If f (z) is analytic inside a circle C with its center at z = a then for all z inside C,

$$f(z)=f(a)+(z-a)\frac{f'(a)}{1!}+(z-a)^2\frac{f''(a)}{2!}+\ldots$$

This is known as Taylors series of f(z) about z = a. This expression is valid at all points interior to any circle. Having its center at z = a and within which the function is analytic.

The circle of convergence of the Taylor series is the largest circle that can be drawn around z = a such that within which f(z) is analytic. The radius of this circle is called the radius of convergence.

Laurent's Series

If C_1 and C_2 are two concentric circles with center at z = a and radii r_1 and r_2 ($r_1 < r_2$) and if f(z) is analytic inside and on the circles and within the annulus between C_1 and C_2 then for any in the annulus we have,

$$f(z)=\sum_{n=1}^{\infty}a_n(z-a)^n+\sum b_n(z-a)^{-n}$$

Where,

$$a_n=\frac{1}{2\pi i}\int_C\frac{f(z)dz}{(z-a)^{n+1}} \text{ and } b_n=\frac{1}{2\pi i}\int_C\frac{f(z)dz}{(z-a)^{-n+1}}$$

Where C is any circle lying between C_1 and C_2 with center at z = a for all n and the integration being taken in positive direction.

Merits and Demerits of Taylor Series

- It can be of high order.
- It is explicit.
- It is easy to show that global error is the same order as LTE.

In the differential equation dy/dx = f(x, y), the function f(x, y) may have a complicated algebraic structure. Then the evaluation of higher order derivatives may become tedious.

Problems

1. Let us determine $y(1.1)$ if $y = x + y, y(1) = 0$ by Taylor series method.

Solution:

Given:

$$y = x + y, y(1) = 0$$

Formula to be used:

$$y(1.1) = y_0 + \frac{h}{1!}y'_0 + \frac{h^2}{2!}y'''_0 + \frac{h^3}{3!}y'''_0$$

$$y = x + y;\ x_0 = 1;\ y_0 = 0;\ x_1 = 1.1 \text{ and } h = 0.1$$

$$y = x + y;\ y'_0 = x_0 + y_0 = 1$$

$$y' = 1 + y;\ y''_0 = 1 + y'_0 = 2$$

$$y'' = y;\ y''_0 = y'''_0 = 2$$

Taylor Series

$$y(1.1) = y_0 + \frac{h}{1!}y'_0 + \frac{h^2}{2!}y'''_0 + \frac{h^3}{3!}y'''_0$$

$$= 0 + (0.1)(1) + \left[(0.1)^2(2)\right]/(2) + \left[(0.1)^3(2)/6\right] + \ldots$$

$$Y(1.1) = 0.1103.$$

2. Let us determine y(0.1) if $\frac{dy}{dx} = 1 + y, y(0) = 1$ using Taylor series method.

Solution:

Given:

$$y' = 1 + y;\ x_0 = 0;\ y_0 = 1;\ x_1 = 0.1;\ h = 0.1$$

Formula to be used:

$$y_1 = y_0 + \frac{h}{1!}y_0 + \frac{h^2}{2!}y_0 + \ldots$$

$$y_1 = y_0 + \frac{h}{1!}y_0 + \frac{h^2}{2!}y_0 + \ldots \qquad \ldots(1)$$

$$Y' = 1 + y;\ y_o = 1 + y_o = 2$$

$$Y'' = y';\ \ y''_o = y_o = 2$$

$$Y''' = y''\ ;\ y'''_o\ y''_o = 2$$

$$y_1 = 1 + (0.1)(2) + \frac{(0.1)^2(2)}{2} + \frac{(0.1)^3(2)}{6} + \ldots$$

$$Y(0.1) = 1 + 0.2 + 0.01 + 0.0003$$

$$= 1.2103.$$

3. Using Taylor series method, let us determine $y(0.1)$ if $y = x^2 + y^2$, $y(0) = 1$.

Solution:

Given:

$$y = x^2 + y^2;\ x_o = 0;\ y_o = 1;\ x = 0.1.$$

$$h = x_1 - x_o = 0.1$$

Formula to be used:

$$y(x) = y_o + \frac{h}{1!}y_o + \frac{h^2}{2!}y''_o + \frac{h^3}{2!}y'''_o + \ldots.$$

$$y' = x^2 + y^2;\ y'_o = x_o^2 + y_o^2 = 0 + 1 = 1$$

$$y'' = 2x + 2y,\ y'' = 2x_o + 2y_o\ y'_o = 0 + 2(1)(1) = 2$$

$$y'' = 2 + 2\left[yy + (y)^2\right];y_o = 8$$

$$y(0{,}1) = 1 + \frac{(0{,}1)}{1}(1) + \frac{(0{,}1)^2}{2}(2) + \frac{(0{,}1)^3}{6}(8)$$

$$= 1 + 0.1 + 0.01 + 0.0013$$

$$y(0{,}1) = 1.113.$$

4. Using Taylor's series method, let us determine y at x = 1.1 by solving the equation

$\frac{dy}{dx} = x^2 + y^2;\ y(1) = 2$. Also let us carryout the computations up to fourth order derivative.

Solution:

Given

$$y' = x^2 + y^2;\ y(1) = 2.$$

$$x_0 = 1;\ y_0 = 2.$$

Formula to be used:

$$y_1 = y_0 + \frac{h}{1!}y'_0 + \frac{h^2}{2!}y''_0 + \frac{h^3}{3!}y'''_0 + \frac{h^4}{4!}y_0^{\prime V} + \ldots$$

$$\therefore x_1 = 1.1;\ h = 0.1.$$

$$y' = x^2 + y^2, \qquad y_0 = x_0^2 + y_0^2 = 1 + 4 = 5$$

$$y'' = 2x + 2yy', \qquad y''_0 = 2x_0 + 2y_0y'_0 = 2 + 2(2)(5) = 22$$

$$y''' = 2 + 2\left[yy'' + y'^2\right], \qquad y'''_0 = 2 + 2\left[y_0\,y''_0 + y'^2_0\right]$$

$$= 2 + 2\left[(2)(22) + 5^2\right]$$

$$= 140.$$

$$Y'^V = 2[yy''' + y''y' + 2y'y'']; \ y_0^{\prime V} = 2[y_0y'''_0 + 3\,y'_0\ y''_0]$$

$$= 2\left[(2)(140) + 3(5)(22)\right] = 1220.$$

Taylor Series:

$$y(x) = y_0 + \frac{h}{1!}y'_0 + \frac{h^2}{2!}y''_0 + \frac{h^3}{3!}y'''_0 + \frac{h^4}{4!}y_0^{\prime V} + \ldots$$

$$y(1.1) = 2 + (0.1)(5) + \frac{(0.1)^2}{2}(22) + \frac{(0.1)^3}{6}(140) + \frac{(0.1)^4}{24}(1220)$$

$$= 2 + 0.5 + 0.11 + 0.0233 + 0.0051$$

$$y(1.1) = 2.6384$$

5. Let us determine the value of y at x = 0.1, 0.2 given $\frac{dy}{dx} = x^2 - 1$, y(0) = 1, by Taylor's series method up to four terms.

Solution:

Given:

$$y' = x^2 y - 1;\ x_0 = 0;\ y_0 = 1;\ x_1 = 0.1.$$

Formula to be used:

Taylor series formula is,

$$y_1 = y_0 + \frac{h}{1!} y'_0 + \frac{h^2}{2!} y''_0 + \frac{h^3}{3!} y'''_0 + \frac{h^4}{4!} y^{IV}_0 + \ldots$$

$y' = x^2 y - 1 = -1$	$y'_0 = x_0^2\, y_0 - 1 = -1$
$y'' = 2xy + x^2 y'$	$y'_0 = 2x_0 y_0 + x_0^2\, y'_0 = 0$
$y''' = 2y + 4xy' + x^2 y''$	$y'''_0 = 2$
$Y^{IV} = 6y'' + 6xy'' + x^2 6'''$	$y^{IV}_0 = -6$

$$h = x_1 - x_0 = 0.1$$

$$y(x_1) = 1 + (0.1)(-1) + \frac{(0.1)^2}{2}(0) + \frac{(0.1)^3}{6}(2) + \frac{(0.1)^4}{24}(..6) + \ldots$$

$$y(0.1) = 0.900305.$$

To find y(0.2) here h = 0.2,

$$y(0.2) = 1 + (0.2)(-1) + \frac{(0.2)^2}{2}(0) + \frac{(0.2)^3}{6}(2) + \frac{(0.2)^4(-6)}{24} + \ldots$$

$$y(0.2) = 0.802266$$

6. Let us use Taylor's formula to expand the function defined by $f(x, y) = x^3 + y^3 + xy^2$ in powers of (x-1) and (y-2).

Solution:

Given:

$$f(x, y) = x^3 + y^3 + xy^2$$

Formula to be used:

$$f(x,y) = f(a,b) + \frac{1}{1!}\left[(x-a)f_x(a,b) + (y-b)f(a,b)\right]$$

$$+\frac{1}{2!}\left[(x-a)^2 f_{xx}(a,b) + 2(x-a)(y-b)f_{xy}(a,b) + (y-b)^2 f_{yy}(a,b)\right]$$

$$+\frac{1}{3!}\left[(x-a)^3 f_{xxx}(a,b) + 3(x-a)^2(y-b)f_{xxy}(a,b)\right.$$

$$\left.+3(x-a)(y-b)^2 f_{xyy}(a,b) + (y-b)^3 f(a,b)\right] + \ldots$$

Here,

$$a = 1,\ b = 2$$

Function	Value at (1,2)
$f(x,y) = x^3 + y^3 + xy^2$	$f(1,2) = 13$
$f_x = 3x^2 + y^2$	$f,(1,2) = 7$
$fy = 3y^2 + 2xy$	$fy(1,2) = 16$
$f_x = 6x$	$f_{xx}(1,2) = 6$
$f_{xy} = 2y$	$f_{xx}(1,2) = 4$
$f_{yy} = 6y + 2x$	$f_{yy}(1,2) = 14$
$f_{xx} = 6$	$f_{xxx}(1,2) = 6$
$f^J_{xxy} = 0$	$f_{xxy}(1,2) = 0$
$f_{xyy} = 2$	$f_{xxy}(1,2) = 2$
$f_{yyy} = 6$	$f_{yyy}(1,2) = 6$

Using these values in (1), we get,

$$x^3 + y^3 + xy^2 = 13 + \frac{1}{1!}\left[(x-1)6 + (y-2)16\right]$$

$$+\frac{1}{2!}\left[(x-1)^2 6+2(x-1)(y-2)4+(y-2)^2 14\right]$$

$$+\frac{1}{3!}\left[(x-1)^3 6+3(x-1)^2 (y-2)^2 2+(y-2)^3 6\right]+...$$

$$=13+6(x-1)+16(y-2)$$

$$+\frac{2}{2}\left[3(x-1)^2+4(x-1)(y-2)+7(y-2)^2\right]$$

$$+\frac{1}{6}\times 6\left[(x-1)^3+(x-1)(y-2)^2+7(y-2)^3\right]+...$$

$$=13+6(x-1)+16(y-2)+3(x-1)^2+4(x-1)(y-2)$$

$$+7(y-2)^2+(x-1)^3+(x-1)(y-2)^2+(y-2)^2+...$$

$$=13+6(x-1)+16(y-2)+3(x-1)^2+4(x-1)(y-2)$$

$$+7(y-2)^2+(x-1)^3+(y-2)^3+(x-1)(y-2)^2+...$$

$$\therefore x^3+y^3+xy^2=13+6(x-1)+16(y-2)+3(x-1)^2+7(y-2)^2+...$$

$$+4(x-1)(y-2)+(x-1)^3+(y-2)^3+(x-1)(y-2)^2+...$$

7. Let us determine the Taylor's series equation ex sin y at the point $\left(-1,\frac{ð}{4}\right)$ up to the third degree terms.

Solution:

Given:

e^x sin y at the point $\left(-1,\frac{\pi}{4}\right)$

Formula to be used:

$$f(x,y)=f(a,b)+\frac{1}{1!}\left[(x-a)f_x(a,b)+(y-b)f_y(a,b)\right]$$

$$+\frac{1}{2!}\left[\begin{matrix}(x-a)^2 f_{xx}(a,b)+2(x-a)(y-b)f_{xy}(a,b)\\+(y-b)^2 f_{yy}(a,b)\end{matrix}\right]$$

$$+\frac{1}{3!}\begin{bmatrix}(x-a)^3 f_{xxx}(a,b)+3(x-a)^2(y-b)f_{xxy}(a,b)\\ +3(x-a)(y-b)^2 f_{xyy}(a,b)+(y-b)^3 f_{yyy}(a,b)\end{bmatrix}+\ldots$$

Function	Value at $\left(-1,\frac{\pi}{4}\right)$
$f(x,y)=e^x \sin y$	$f=\frac{1}{e}\left(\frac{1}{\sqrt{2}}\right)=\frac{1}{e\sqrt{2}}$
$f_x=e^x \sin y$	$f_x=\frac{1}{e}\left(\frac{1}{\sqrt{2}}\right)=\frac{1}{e\sqrt{2}}$
$f_y=e^x \cos y$	$f_y=\frac{1}{e}\left(\frac{1}{\sqrt{2}}\right)=\frac{1}{e\sqrt{2}}$
$f_{xx}=e^x \sin y$	$f_{xx}=\frac{1}{e}\left(\frac{1}{\sqrt{2}}\right)=\frac{1}{e\sqrt{2}}$
$f_{yy}=-e^x \sin y$	$f_{yy}=-\frac{1}{e}\left(\frac{1}{\sqrt{2}}\right)=-\frac{1}{e\sqrt{2}}$
$f_{xxx}=e^x \sin y$	$f_{xxx}=\frac{1}{e}\left(\frac{1}{\sqrt{2}}\right)=\frac{1}{e\sqrt{2}}$
$f_{xxy}=e^x \cos y$	$f_{xxy}=\frac{1}{e}\left(\frac{1}{\sqrt{2}}\right)=\frac{1}{e\sqrt{2}}$
$f_{xyy}=-e^x \sin y$	$f_{xxy}=-\frac{1}{e}\left(\frac{1}{\sqrt{2}}\right)=-\frac{1}{e\sqrt{2}}$
$f_{yyy}=-e^x \cos y$	$f_{yyy}=-\frac{1}{e}\left(\frac{1}{\sqrt{2}}\right)=-\frac{1}{e\sqrt{2}}$

$$f(x,y)=f(a,b)+\frac{1}{1!}\left[(x-a)f_x(a,b)+(y-b)f_y(a,b)\right]$$

$$+\frac{1}{2!}\begin{bmatrix}(x-a)^2 f_{xx}(a,b)+2(x-a)(y-b)f_{xy}(a,b)\\ +(y-b)^2 f_{yy}(a,b)\end{bmatrix}$$

$$+\frac{1}{3!}\begin{bmatrix}(x-a)^3 f_{xxx}(a,b)+3(x-a)^2(y-b)f_{xxy}(a,b)\\ +3(x-a)(y-b)^2 f_{xyy}(a,b)+(y-b)^3 f_{yyy}(a,b)\end{bmatrix}+\ldots$$

Here,

$$a=-1, b=\frac{\pi}{4}; x-a=x+1;\ y-b=y-\frac{\pi}{4}$$

$$f(x,y)=\frac{1}{e}\frac{1}{\sqrt{2}}+\left[(x+1)\frac{1}{e}\frac{1}{\sqrt{2}}+\left(y-\frac{\pi}{4}\right)\frac{1}{e}\frac{1}{\sqrt{2}}\right]$$

$$+\frac{1}{2!}\left[(x+1)^2\frac{1}{e}\frac{1}{\sqrt{2}}+2(x+1)\left(y-\frac{\pi}{4}\right)\left(\frac{-1}{e}\frac{1}{\sqrt{2}}\right)+\left(y-\frac{\pi}{4}\right)^2\frac{1}{e}\frac{1}{\sqrt{2}}\right]$$

$$+\frac{1}{3!}\left[(x+1)^3\frac{1}{e}\frac{1}{\sqrt{2}}+3(x+1)^2\left(y-\frac{\pi}{4}\right)\frac{1}{e}\frac{1}{\sqrt{2}}+3(x+1)\left(y-\frac{\pi}{4}\right)^2\left(\frac{-1}{e}\frac{1}{\sqrt{2}}\right)\right]+\dots$$

8. Let us determine the Taylor's series equation $x^2y^2+2x^2y+3xy^2$ in powers of (x + 2) and (y -1) up to 3rd degree terms.

Solution:

Given:

$$f(x,y)=x^2y^2+2x^2y+3xy^2 \qquad ...(1)$$

Formula to be used:

$$f(x,y)=f(a,b)+\frac{1}{1!}\left[(x-a)f_x(a,b)+(y-b)f_y(a,b)\right]$$

$$+\frac{1}{2!}\left[\begin{array}{l}(x-a)^2 f_{xx}(a,b)+2(x-a)(y-b)f_{xy}(a,b)\\ +(y-b)^2 f_{yy}(a,b)\end{array}\right]$$

$$+\frac{1}{3!}\left[\begin{array}{l}(x-a)^3 f_{xxx}(a,b)+3(x-a)^2(y-b)f_{xxy}(a,b)\\ +3(x-a)(y-b)^2 f_{xyy}(a,b)+(y-b)^3 f_{yyy}(a,b)\end{array}\right]+\dots$$

We need Taylor series in powers of (x + 2) and (y – 1).

i.e., at the point x = -2 and y = 1,

$$f(x,y)=x^2y^2+\ 2x^2y^2+3xy^2$$

$$f(-2,\ 1)=(-2)^2(1)^2+2\ (-2)^2(1)+3(-2)(1)^2\ =6$$

$$f_x=2xy^2+4xy+3y^2$$

$$f_x(-2,1)=2(-2)(1^2)+4(-2)(1)+3\ (1^2)=-9$$

$$fy = 2x^2y + 2x^2 + 6xy$$

$$_y(-2,1) = 2(-2)^2\ (1) + 2(-2)^2 + 6(-2)(1) = 8 + 8 - 12 = 4$$

$$f_{xx} = 2y^2 + 4y$$

$$f_{xy}(-2,1) = 2\ (1)^2 + 4(1) = 6$$

$$f_{xy} = 4xy + 4x + 6y$$

$$f_{xy}(-2,1) = 4(-2)(1) + 4\ (-2) + 6\ (1) = -\ 10$$

$$f_{yy} = 2x^2 + 6x$$

$$f_{yy}(-2,1) = 2(-2)^2 + 6\ (-2) = 8 - 12 = -4$$

$$f_{xxx} = 0.$$

$$f_{xxx}(-2,1) = 0$$

$$f_{xxy} = 4y + 4$$

$$f\ _{xxy}(-2,1) = 4(1) + 4 = 8$$

$$f_{xyy} = 4x + 6$$

$$f_{xyy} = 4(-2) + 6 = -2$$

$$f_{yyy} = 0$$

$$f_{yyy}(-2,1) = 0$$

Taylor's series of f(x),

$$f(x,y) = f(-2,1) + \frac{1}{1!}\left[(x+2)f_x + (y-1)f_y\right]$$

$$+\ 1/2!\left[(x+2)^2\ f_{xx} + 2\ (x+2)(y-1)\ f_{xy} + (y-1)^2\ f_{yy}\right]$$

$$+\frac{1}{3!}\left[(x+2^3)f_{xxx} + 3(x+2)^2(y-1)f_{xxy} + 3(x+2)(y-1)^3 f_{yyy}\right] + \ldots$$

$$f(x,y)=6+1/1!\left[\left(-9(x+2)+4(y-1)\right)\right]+1/2!\left[6(x+2)^2-20\ (x+2)(y-1)-4\ (y-1)^2\right]$$

$$+\ 1/3!\left[24(x+2)^2(y-1)-6(x+2)(y-1)^2\right]+\ \ldots$$

9. Let us expand the Taylor's equation ex log (1+y) in powers of x and y up to the third degree terms.

Solution:

Given:

$$e^x\ \log\ (1+y)$$

Formula to be used:

$$f(x,y)=f(a,b)+\frac{1}{1!}\left[(x-a)f_x(a,b)+(y-b)f_y(a,b)\right]$$

$$+\frac{1}{2!}\begin{bmatrix}(x-a)^2 f_{xx}(a,b)+2(x-a)(y-b)f_{xy}(a,b)\\+(y-b)^2 f_{yy}(a,b)\end{bmatrix}$$

$$+\frac{1}{3!}\begin{bmatrix}(x-a)^3 f_{xxx}(a,b)+3(x-a)^2(y-b)f_{xxy}(a,b)\\+3(x-a)(y-b)^2 f_{xyy}(a,b)+(y-b)^3 f_{yyy}(a,b)\end{bmatrix}+\ldots$$

Function	Value at (0,0)
$f_{(x,y)}=e^x\log(1+y)$	$f=0$
$f_{(x,y)}=e^x\log(1+y)$	$f_x=0$
$f_y=e^x(1+y)^{-1}$	$f_y=1$
$f_{xx}=e^x\log(1+y)$	$f_{xx}=0$
$f_{xx}=e^x\log(1+y)$	$f_{xy}=1$
$f_{yy}=-e^x(1+y)^{-2}$	$f_{yy}=-1$
$f_{xxx}=e^x\log(1+y)$	$f_{xxx}=0$
$f_{xxy}=-e^x(1+y)^{-1}$	$f_{xxy}=1$

$f_{xyy} = 2e^{x} \log(1+y)^{-2}$	$f_{xyy} = -1$
$f_{yyy} = 2e^{x} \log(1+y)^{-3}$	$f_{yyy} = 2$

Applying Taylor's theorem, we get,

$$e^{x} \log(1+y) = 0 + \frac{x(0)+y(1)}{1!} + \frac{x^2(0)+2xy(1)+y^2(-1)}{2!}$$

$$+\frac{x^3(0)+3x^2y(1)+3xy^2(-1)+y^3(2)}{3!}+\ldots$$

$$\frac{y}{1!}+\frac{2xy-y^2}{2!}+\frac{3x^2y-3xy^2+2y^3}{3!}+\ldots$$

10. Let us assume the equation $u = f\left[\frac{x}{y},\frac{y}{z},\frac{z}{k}\right]$. Here we prove that $x\frac{\partial u}{\partial x}+y\frac{\partial u}{\partial y}+z\frac{\partial u}{\partial z}=0$.

Solution:

Given:

$$u = f\left[\frac{x}{y},\frac{y}{z},\frac{z}{k}\right]$$

Given u is a function of x, y, z

And,

$$u(x,y,z) = \left(\frac{x}{y},\frac{y}{z},\frac{z}{x}\right)$$

$$u(tx,ty,tz) = \left(\frac{tx}{ty},\frac{ty}{tz},\frac{tz}{tx}\right)$$

$$= f\left(\frac{x}{y},\frac{y}{z},\frac{z}{x}\right)$$

$\therefore$ u (x, y, z) is a homogeneous function of x, y, z

By Euler's theorem, $x\frac{\partial u}{\partial x}+y\frac{\partial u}{\partial y}+z\frac{\partial u}{\partial z}=0$.

11. Using Taylor's series expansion, let us express $f(x,y)=e^{2x}\cos 2y$ in powers of x and y up to second degree.

Solution:

Given:

$$f(x,y)=e^{2x}\cos 2y \qquad f\left(0,\frac{\pi}{2}\right)=-1$$

Formula to be used:

$$f(x,y)=f(a,b)+\frac{1}{1!}\left[(x-a)f_x(a,b)+(y-b)f_y(a,b)\right]$$

$$+\frac{1}{2!}\begin{bmatrix}(x-a)^2 f_{xx}(a,b)+2(x-a)(y-b)f_{xy}(a,b)\\ +(y-b)^2 f_{yy}(a,b)\end{bmatrix}$$

$$+\frac{1}{3!}\begin{bmatrix}(x-a)^3 f_{xxx}(a,b)+3(x-a)^2(y-b)f_{xxy}(a,b)\\ +3(x-a)(y-b)^2 f_{xyy}(a,b)+(y-b)^3 f_{yyy}(a,b)\end{bmatrix}+\dots$$

$$f_x(x,y)=e^{2x}2\cos 2y \qquad f_x\left(0,\frac{\pi}{2}\right)=-2$$

$$f_{xx}(x,y)=e^{2x}4\cos 2y \qquad f_{xx}\left(0,\frac{\pi}{2}\right)=-4$$

$$f_{xy}(x,y)=-4e^{2x}\sin 2y \qquad f_{xy}\left(0,-\right)=0$$

$$f_y(x,y)=-2e^{2x}\sin 2y \qquad f_y\left(0,\frac{\pi}{2}\right)=0$$

$$f_{yy}(x,y)=-4e^{2x}\cos 2y \qquad f_{yy}\left(0,\frac{\pi}{2}\right)=4$$

Taylor's series is $f(x,y)=-1-2x+2\left(-x^2+\left(y-\frac{\pi}{2}\right)^2\right)+\dots$

12. Let us evaluate $\int e^{-x^2}dx$ as an infinite series and let us evaluate ${}_0\int^1 e^{-x^2}dx$ correct to within an error of 0.001.

Solution:

Given:

$$\int e^{-x^2}dx$$

Error = 0.001

Formula to be used:

$$e^{-x^2}=\sum_{n=0}^{\infty}\frac{\left(-x^2\right)^n}{n!}=\sum_{n=0}^{\infty}(-1)^n\frac{x^{2n}}{n!}=1-\frac{x^2}{1!}+\frac{x^4}{2!}-\frac{x^6}{3!}+\dots$$

For all values of x,

$$e^{-x^2}=\sum_{n=0}^{\infty}\frac{\left(-x^2\right)^n}{n!}=\sum_{n=0}^{\infty}(-1)^n\frac{x^{2n}}{n!}=1-\frac{x^2}{1!}+\frac{x^4}{2!}-\frac{x^6}{3!}+\dots$$

Now we integrate term by term,

$$\int e^{-x^2}dx=\int\left(1-\frac{x^2}{1!}+\frac{x^4}{2!}+\frac{x6}{3!}+\dots+(-1)^n\frac{x^{2n}}{n!}+\dots\right)dx$$

$$=C+x-\frac{x^3}{3\cdot 1!}+\frac{x^5}{5\cdot 2!}-\frac{x^7}{7\cdot 3!}+\dots+(-1)^n\frac{x^{2n+1}}{(2n+1)n!}+\dots$$

This series converges for all x because the original series for converges for all x.

The Fundamental Theorem of Calculus gives,

$$\int_0^1 e^{-x^2}dx=\left[x-\frac{x^3}{3\cdot 1!}+\frac{x^5}{5\cdot 2!}-\frac{x^7}{7\cdot 3!}-\dots\right]_0^1$$

$$=1-\frac{1}{3}+\frac{1}{10}-\frac{1}{42}+\frac{1}{216}-\dots$$

$$\approx 1-\frac{1}{3}+\frac{1}{10}-\frac{1}{42}+\frac{1}{216}\approx 0.7475$$

The Alternating Series Estimation Theorem shows that the error involved in this approximation is less than $\frac{1}{11\cdot 5!}=\frac{1}{1320}<0.001.$

13. Let us find the Mc Laurent's series for and prove that it represents for all.

Solution:

We arrange our computation in two columns as follows:

$$f(x)=\sin x \qquad f(0)=0$$

$$f'(x)=\cos x \qquad f'(0)=1$$

$$f''(x)=-\sin x \qquad f''(0)=0$$

$$f'''(x)=-\cos x \qquad f'''(0)=-1$$

Since the derivatives repeat in a cycle of four, we can write the Mc Laurent's series as follows:

$$f(0)+\frac{f'(0)}{1!}x+\frac{f''(0)}{2!}x^2+\frac{f'''(0)}{3!}x^3+\frac{f^{(4)}(0)}{4!}x^4+\ldots=0+\frac{1}{1!}x+\frac{0}{2!}x^2+\frac{-1}{3!}x^3+\frac{0}{4!}x^4+\ldots$$

$$=x-\frac{x^3}{3!}+\frac{x^5}{5!}+\frac{x^7}{7!}+\ldots$$

$$=\sum_{n=0}^{\infty}(-1)^n\frac{x^{2n+1}}{(2n+1)!}$$

Using the remainder term (10) with a=0, we have $R_n(x)=\frac{f^{(n+1)}(z)}{(n+1)!}x^{n+1}$ where $f(x)=\sin x$ and lies between 0 and x. But $f^{(n+1)}(z)$ is $\pm\sin z$ or $\pm\cos z$ In any case, $\left|f^{(n+1)}(z)\right|\le 1$ and so,

$$0\le\left|R_{(n)}(x)\right|=\frac{\left|f^{(n+1)}(z)\right|}{(n+1)!}=\left|x^{n+1}\right|\le\frac{1}{(n+1)!}\left|x^{n+1}\right|=\frac{|x|^{n+1}}{(n+1)!}$$

By below equation,

$$\lim_{n\to\infty}\frac{x^n}{n!}=0 \text{ for every real number } x$$

The right side of this inequality approaches 0 as $n\to\infty$, so $R_{(n)}(x)\to 0$ by the Squeeze

Theorem. It follows that $R_{(n)}(x) \to 0$ as, $n \to \infty$ so $\sin x$ is equal to the sum of its Mc Laurent's series by the below theorem,

If $f(x)=T_n(x)+R_n(x)$, where T_n is the nth th-degree Taylor polynomial of f at a and $\lim_{n\to\infty} R_n(x)=0$ for $|x-a|<R$, , then f is equal to the sum of its Taylor series on the interval $|x-a|<R$. Thus

$$\sin x = x-\frac{x^3}{3!}+\frac{x^5}{5!}+\frac{x^7}{7!}+\ldots=\sum_{n=0}^{n}(-1)^n\frac{x^{2n+1}}{(2n+1)!} \text{ for all x}$$

14. Let us find the Mc Laurent's series for and its radius of convergence.

Solution:

Given:

$$f(x)=\frac{1}{\sqrt{4-x}}$$

$$\frac{1}{1-x}=\sum_{n=0}^{\infty}x^n=1+x+x^2+x^3+\ldots \qquad R=1$$

$$e^x=\sum_{n=0}^{\infty}\frac{x^n}{n!}=1+\frac{x}{1!}+\frac{x^2}{2!}+\frac{x^3}{3!}+\ldots \qquad R=\infty$$

$$\sin x=\sum_{n=0}^{\infty}(-1)^n\frac{x^{2n+1}}{(2n+1)!}=x+\frac{x^3}{3!}+\frac{x^5}{5!}+\frac{x^7}{7!}+\ldots \qquad R=\infty$$

$$\cos x=\sum_{n=0}^{\infty}(-1)^n\frac{x^{2n}}{(2n)!}=1-\frac{x^2}{2!}+\frac{x^4}{4!}+\frac{x^6}{6!}+\ldots \qquad R=\infty$$

$$\tan^{-1}x=\sum_{n=0}^{\infty}(-1)^n\frac{x^{2n+1}}{2n+1}=x-\frac{x^3}{3}-\frac{x^5}{5}+\frac{x^7}{7}+\ldots \qquad R=1$$

$$\ln(1+x)=\sum_{n=0}^{\infty}(-1)^{n-1}\frac{x^n}{n}=x-\frac{x^2}{2}+\frac{x^3}{3}-\frac{x^4}{4}+\ldots \qquad R=1$$

$$(1+x)^k=\sum_{n=0}^{\infty}\binom{k}{n}x^n=1+kn+\frac{k(k-1)}{2!}x^2+\frac{k(k-1)(k-2)}{3!}x^3+\ldots \qquad R=1$$

One reason that Taylor series are important is that they enable us to integrate functions that we couldn't previously handle. In fact, the function can't be integrated by techniques discussed so far because its ant derivative is not an elementary function.

15. Let us evaluate $\lim_{x \to 0} \frac{e^x - 1 - x}{x^2}$.

Solution:

Given:

$$\lim_{x \to 0} \frac{e^x - 1 - x}{x^2}$$

Using the Mc Laurent's series for e^x, we have,

$$\lim_{x \to 0} \frac{e^x - 1 - x}{x^2} = \lim_{x \to 0} \frac{\left(1 + \frac{x}{1!} + \frac{x^2}{2!} + \frac{x^3}{3!} + \ldots\right) - 1 - x}{x^2}$$

$$= \lim_{x \to 0} \frac{\frac{x^2}{2!} + \frac{x^3}{3!} + \frac{x^4}{4!} + \ldots}{x^2}$$

$$= \lim_{x \to 0} \left(\frac{1}{2} + \frac{x}{3!} + \frac{x^2}{4!} + \frac{x^3}{5!} + \ldots\right) = \frac{1}{2}$$

Because power series are continuous functions.

4.2 Singularities and Zeros

A singularity is a point at which an equation, surface, etc., blows up or becomes degenerate. Singularities are often also called singular points.

Singularities are extremely important in complex analysis where they characterize the possible behaviors of analytic functions. Complex singularities are points in the domain of a function f where fails to be analytic. Isolated singularities may be classified as poles, essential singularities, logarithm f_{ic} singularities or removable singularities. Non isolated singularities may arise as natural boundaries or branch cuts.

Singularities

Zeros of an Analytic Function

If a function f(z) analytic in a region R is zero at a point $z = z_0$ in R then z_0 is called a zero of f(z).

Simple zero:

- If $f(z_0) = 0$ and $f'(z_0)$ not equal to 0 then $z = z_0$ is called a simple zero of f(z) or a zero of the first order.

Zero of order n:

- An analytic function f(z) is said to have a zero of order n if f(z) can be expressed as $f(z) = (z - z_0)^m \varphi(z)$ where $\varphi(z)$ is analytic and $\varphi(zo)$ not equal to 0.

Singular points:

- A point $z = z_0$ at which a function f(z) fails to be analytic is called a singular point.

Entire Function:

- A function f(z) which is analytic everywhere in the finite plane is called an entire function.

Meromorphic function:

- A function f(z) which is analytic everywhere in the finite plane except at finite number of poles is called a meromorphic function.

Types of Singularities

Isolated Singularity:

A point $z = z_0$ is said to be isolated singularity of f(z) if,

- f(z) is not analytic at $z = z_0$.
- There exist a neighborhood of $z = z_0$ containing no other singularity.

Removable Singularity: If the principal part of f(z) in Laurent series expansion of f(z) about the point z_0 is zero then the point $z = z_0$ is called removable singularity.

Pole: If we can find a positive integer n such that $\lim_{z \to a}(z - a)^n f(z)$not equal to 0 then z = a is called a pole of order n for f(z).

Problems

1. Let us determine the zeros of $z^3 - 1/z^3 + 1$.

Solution:

Given:

$$z^3 - 1/z^3 + 1$$

The solutions of f(z) are given by f(z) = 0.

That is $z^3 - 1 = 0$

$$z = 1,\ \omega,\ \omega^2.$$

2. Let us determine the zeros of $\frac{\sin z - z}{z^3}$.

Solution:

The solutions of f(z) are given by f(z) = 0.

That is,

$$f(z) = \frac{\sin z - z}{z^3}$$

$$= \frac{\frac{-z^3}{3!} + \frac{-z^5}{5!} - \dots}{z^3}$$

$$= \frac{-1}{3!} + \frac{z^2}{5!} - \dots$$

Now,

$$\lim_{z \to 0} \frac{\sin z - 2}{z^3} \neq 0$$

3. Let us determine the nature of singularity z = 0 of the function.

Solution:

Given:

The function f(z) is not defined at z=0.

By L'Hospital rule

$$\lim_{z \to 0} \frac{\sin z - z}{z^3} = \lim_{z \to 0} \frac{\cos z - 1}{3z^2}$$

$$= \lim_{z \to 0} \frac{-\sin z}{6z}$$

$$= \frac{-1}{6}$$

Since the limit exist and is finite, the singularity at z = 0 is a removable singularity.

4. Let us determine the nature of singularity at z = 0 of $f(z) = \sin z / z$.

Solution:

Given:

$$f(z) = \sin z / z$$

The function f(z) is not defined at z = 0.

By L'Hospital rule,

$$\lim_{x \to 0} \frac{\sin z}{z} = \lim_{x \to 0} \frac{\cos z}{1}$$

Since the limit exist and is finite, the singularity at z = 0 is a removable singularity.

5. Let us classify the singularity of the function $f(z) = Z - 2/z^2 \sin(1/z - 1)$.

Solution:

Poles of f(z) are z = 0,0.That is z = 0 is a pole of order 2.

Zeros of f(z) are obtained by,(z - 2)sin1/(z - 1) = 0.

$$\Rightarrow z - 2 = 0 \quad \text{and } \sin\left(\frac{1}{z-1}\right) = 0$$

$$\Rightarrow z - 2 = 0 \text{ and } z = \frac{1}{n\pi} + 1, n = 0,1,2,....$$

The limits of zeros is 1. Therefore, z = 1 is isolated and essential singularity.

6. Let us evaluate the following problem using removable singularities.

Solution:

Let,

$$f(z) = \frac{\sin z}{z}$$

Clearly $z = 0$ is a singular point,

$$f(z) = \frac{\sin z}{z} = \frac{1}{z}\left[z - \frac{z^2}{3!} + \frac{z^5}{5!} - ....\right]$$

$$= 1 - \frac{z^2}{3!} + \frac{z^4}{5!} - \ldots$$

$z = 0$ is a removable singular point.

4.3 Residue Integration Method

Residue is a complex number proportional to the contour integral of a meromorphic function along a path enclosing one of its singularities.

The residue of a meromorphic function f at an isolated singularity a, often denoted $\mathrm{Res}_a(f, a)$ or $\mathrm{Res}_a(f)$, is the unique value R such that f(z) - R/(z - a) has an analytic anti-derivative in a punctured disk $0 < |z - a| < \delta$.

Definition

If $z = z_0$ is an isolated singular point of f (z) we can find the Laurent's series of f(z)about z = zo

$$f(z) = \sum_{n=0}^{\infty} a_n (z - z_0)^n + \sum_{n=1}^{\infty} \frac{b_n}{(z - z_0)^n}$$

The coefficient of $\frac{1}{(z - z_0)}$ in the above expansion is called the residue of f(z) at $Z = z_0$.

Evaluation of Residues

Suppose z = a is a pole of order 1

Then $\{\mathrm{Res} f(z)\}_{z=a} = \lim_{z \to a} \frac{1}{(n-1)!} \frac{d^{n-1}}{dz^{n-1}} (z - a)^n f(z)$ Suppose z = a is a pole of order n

Then $\{\mathrm{Res} f(z)\}_{z=a} = \lim_{z \to a} (z - a) f(z)$ Suppose z = a is $f(z) = \frac{P(z)}{Q(z)}$ a pole of order 1 and

Then $\{\mathrm{Res} f(z)\}_{z=a} = P(a)/Q'(a)$.

Cauchy's Residue Theorem

If f(z) is analytic at all point inside and on a simple closed curve C except at a finite number of point $z_1, z_2, z_3, \ldots z_n$ inside C then

$$\int_c f(z)\, dz = 2\pi i \text{ [Sum of residues of f(z) at } z_1, z_2, z_3, \ldots z_n]]$$

Proof:

Given that f(z) is not analytic only at $Z_1, Z_2, Z_3, ... Z_n$.

Draw the non-intersecting small circles c_1, c_2, c_3.....c_n with center at Z_1, Z_2, Z_3,...Z_n and radii. Then f(z) is analytic in the region between c and c_1, c_2, c_3.....c_n

$$\int_c f(z)dz = \int_{c_1} f(z)dz + \int_{c_2} f(z)dz + ... + \int_{c_n} f(z)dz\, f(z)dz \qquad ...(1)$$

Now $z_1, z_2, z_3, ...z_n$ are the singular points of $f(z)$

$\therefore$Res $f(z)_{x=z_i}$ = the coefficient of $\frac{1}{z-z_i}$ in the Laurent's series of $f(z)$ about $z-z_i$ (by definition of residues)

$$= b_1 = \frac{1}{2\pi i}\int_{c_1} \frac{f(z)}{(z-z_i)^{1-n}} dz$$

Since,

$$\left(b_n = \frac{1}{2\pi i}\int_{c_1} \frac{f(z)}{(z-z_i)^{z-n}} dz \right)$$

$$= \frac{1}{2\pi i}\int_{c_1} \frac{f(z)}{(z-z_i)^{o}} dz$$

$$= \frac{1}{2\pi i}\int_{c_1} f(z)dz$$

$$\Rightarrow \int_{c_1} f(z)dz = 2\pi i \,|\, \text{Res} f(z) |_{z=z_i} \qquad ...(2)$$

From (1) and (2),

$$\int_{c_1} f(z)dz = 2\pi \,|\, \text{Res} f(z) |_{z=z_1} + 2\pi \,|\, \text{Res} f(z) |_{z=z_2} + 2\pi \,|\, \text{Res} f(z) |_{z=z_n}$$

$$= 2\pi \,|\, \text{Res} f(z) |_{z=z_1} + |\, \text{Res} f(z) |_{z=z_2} + ... + |\, \text{Res} f(z) |_{z=z_n}$$

$2\pi i$ {Sum of residuals of $f(z)$ at $z = z_1, z_2, z_3.......z_n$}.

1. Let us calculate the residue of $f(z)=1-e^{2x}/z^3$.

Solution:

Given:

$$f(z)=1-e^{2x}/z^3$$

Here,

z = 0 is a pole of order 3.

$$\text{Res at } z=0=\frac{1}{2!}\lim_{x\to 0}\frac{d^2}{dz^2}\left[z^3\frac{1-e^{2x}}{z^3}\right]$$

$$=\frac{1}{2!}\lim_{z\to 0}\frac{d}{dz}\left[-2e^{2x}\right]$$

$$=\frac{1}{2!}\lim_{z\to 0}\frac{d}{dz}\left[-4e^{2x}\right]$$

$$=-2.$$

2. Let us determine the residue of $f(z)=z/(z-1)^2$ at its pole.

Solution:

Given:

$$f(z)=z/(z-1)^2$$

Here,

z = 0 is a pole of order 2.

$$\text{Res at } (z=1)=\frac{1}{1!}\lim_{x\to 1}\frac{d}{dz}\left[(z-1)^2\frac{z}{(z-1)^2}\right]$$

$$=\lim_{x\to 1}\frac{d}{dz}z$$

$$=1.$$

3. Let us evaluate $\int_C \frac{z-1}{(z+1)^2(z-2)}dz$, where C is the circle $|z-i|=2$ using Cauchy's residue theorem.

Solution:

Let,

$$I = \int_C \frac{z-1}{(z+1)^2(z-2)} dz$$

$$f(z) = \frac{z-1}{(z+1)^2(z-2)}.$$

z = 2 is a simple pole, lies outside C,

$$\left[\because 12 - i1 = \sqrt{4+1} = \sqrt{5} > 2\right]$$

z = − 1 is a double pole, lies inside C,

$$\left[\because 1 - 1 = i1\sqrt{2} < 2\right].$$

WKT $\int_c f(z)dz = 2\pi i$ [Sum of Residues of f(z) at its poles].

$$\left[\text{Res of } f(z)\right]_z = 2 = 0.$$

$$\left[\text{Res of } f(z)\right]_z = -1 = \underset{z\to -1}{Lt} \frac{d}{dz}\left[(z+1)^2 f(z)\right].$$

$$= \underset{z\to -1}{Lt} \frac{d}{dz}\left[(z+1)^2 \frac{z-1}{(z+1)^2(z-2)}\right]$$

$$= \underset{z\to -1}{Lt} \frac{d}{dz}\left[\frac{z-1}{z-2}\right]$$

$$= \underset{z\to -1}{Lt}\left[\frac{(z-2)(1)-(z-1)(1)}{(z-2)^2}\right]$$

$$= \underset{z\to -1}{Lt}\left[\frac{-1}{(z-2)^2}\right] = \frac{-1}{(-1-2)^2} = \frac{-1}{9}.$$

$$\therefore I = \int_c \frac{z-1}{(z+1)^2(z-2)} dz$$

$$\int f(z)dz$$

$$= 2\pi\left[0+\left(-\tfrac{1}{9}\right)\right] = \frac{-2\pi i}{9}.$$

4. Let us evaluate $\int_0^\infty \frac{x^2 dx}{(x^2+a^2)(x^2+b^2)}, a>0, b>0.$

Solution:

Consider $\int_0^\infty \frac{x^2 dx}{(x^2+a^2)(x^2+b^2)}, a>0, b>0.$ and choose the contour C consisting of the upper semi-circle r, |z | = R together with the diameter (– R, R) on real axis,

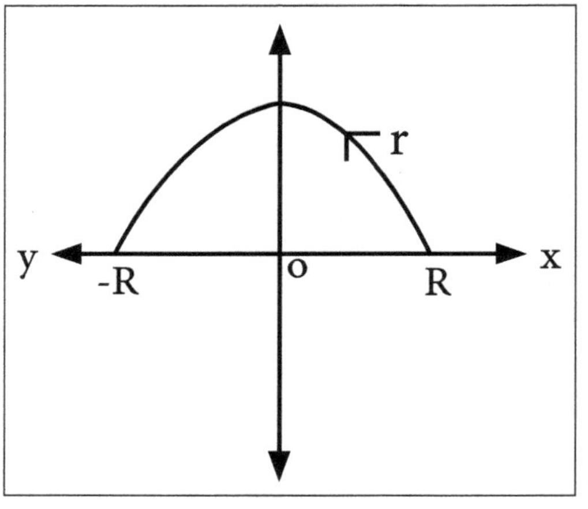

The singularities of f(z) are Z = ± a_i – and Z = ± b_i –. Here z = a_i and b_i lies inside the curve C: and z = – a_i and – b_i lies outside the curve C.

Consider,

$$\int_C f(z)dz = 2\pi i\left[\operatorname*{Res}_{z=ai} f(z) + \operatorname*{Res}_{z=bi} f(z)\right] \quad ...(1)$$

$$f(z) = \frac{z^2}{(z^2+a^2)(z^2+b^2)}$$

$$\operatorname*{Res}_{z\to ai}(f(z)) = \lim_{z\to ai}(z-ai)f(z)$$

$$= \lim_{z\to ai}(z-ai)\frac{z^2}{(z+ai)(z-ai)(z^2+b^2)}$$

$$= \frac{-a^2}{2ai\left(-a^2+b^2\right)} = \frac{a}{2i\left(a^2-b^2\right)}$$

$$\operatorname*{Res}_{z\to bi}\left(f(z)\right) = \lim_{z\to bi}(z-bi)f(z)$$

$$= \lim_{z\to bi}(z-bi)\frac{z^2}{\left(z^2+a^2\right)(z+bi)(z-bi)}$$

$$= \frac{-b^2}{\left(a^2-b^2\right)2bi} = \frac{-b}{2i\left(a^2-b^2\right)}.$$

Substitute in (1),

$$\int_c f(z)dz = 2\pi i\left[\frac{a}{2i\left(a^2-b^2\right)} - \frac{b}{2i\left(a^2-b^2\right)}\right]$$

$$= \frac{\pi}{(a+b)(a-b)}(a-b) = \frac{\pi}{a+b} \qquad ...(2)$$

We know that,

$$\int_c f(z)dz = \int_F f(z)dz + \int_{-R}^{R} f(x)dx$$

As $R \to \infty$ then $\int \to 0$ by Cauchy's lemma.

$$\therefore \int_C f(z)dz = \int_{-\infty}^{\infty} f(x)dx$$

i.e., $$\int_{-\infty}^{\infty}\frac{x^2dx}{\left(x^2+a^2\right)\left(x^2+b^2\right)} = \int_c \frac{z^2dz}{\left(z^2+a^2\right)\left(z^2+b^2\right)} = \frac{\pi}{a+b} \qquad ...(3)$$

Here f(x) is even $\int_{-\infty}^{\infty} f(x)dx = 2\int_0^{\infty} f(x)\,dx$

$$2\int_0^{\infty}\frac{x^2dx}{\left(x^2+a^2\right)\left(x^2+b^2\right)} = \frac{\pi}{a+b}$$

$$\Rightarrow \int_0^{\infty}\frac{x^2dx}{\left(x^2+a^2\right)\left(x^2+b^2\right)} = \frac{\pi}{2(a+b)}.$$

4.3.1 Evaluation of Real Integrals

Residue Theorem

Let Ω be a simply connected domain and A be an isolated subset of Ω.

Suppose f: Ω\A →C is a holomorphic function. The for any simple closed curve γ in Ω \ A, we have,

$$\int_{\gamma} f(z)dz = 2\pi\iota \sum_{a\in A} R_a(f)\eta(\gamma;a)$$

where η(γ; a) denotes the winding number of γ around a.

Proof:

If R is the region in Ω enclosed by γ note that R ∪ γ is a closed and bounded subset of C. Since A is an isolated set, it follows that A∩R = A∩(R ∪ γ) has only finitely many elements say $a_1, ..., a_k$. Choose r > 0 so that the disc of radius r with center a_j are all contained in Ω and are all mutually disjoint and disjoint form γ. Let C_j denote the positively oriented boundary of this disc. Then by II version of Cauchy's theorem, we have,

$$\int_{\gamma} f(z)dz = \sum_{j=1}^{k} \int_{c_j} f(z) = 2\pi\iota \sum_{j=1}^{k} R_{a_j}(f)$$

Since η(γ, a) = 0 if a ≠ a_j and is equal to 1 if a = a_j, the conclusion of the theorem.

TYPE I

Integrals of the type $\int_0^{2\pi} f(\cos\theta, \sin\theta)d\theta$ where f(cos θ, sin θ) is a rational function of sin θ and cos θ. In this case we take unit circle |z|= 1 as the contour. On |z|=1.

$$z = e^{i\theta}, \cos\theta = \frac{1}{2}\left(e^{i\theta} + e^{-i\theta}\right) = \frac{1}{2}\left(z + \frac{1}{z}\right)$$

$$dz = ie^{i\theta}d\theta, \sin\theta = \frac{1}{2i}\left(e^{i\theta} - e^{-i\theta}\right) = \frac{1}{2i}\left(z - \frac{1}{z}\right)$$

$$\Rightarrow d\theta = \frac{dz}{iz}$$

Also varies from 0 to 2π =

$$\therefore \int_0^{2\pi} f(\cos\theta, \sin\theta)d\theta = \int_c f\left(\frac{z^2+1}{2z}, \frac{z^2-1}{2iz}\right)\frac{dz}{iz}$$

Now applying Cauchy's residue theorem, we can evaluate the Integral on the right side.

TYPE II

Integration Around Semi-circular Contour

Consider the integral,

Improper integrals of the form where, $\int_{-\infty}^{\infty} \frac{P(x)}{Q(x)} dx.$ P(x) and Q(x)

Are polynomials in x such that the degree of Q exceeds that of P at least by two and Q(x) does not vanish for any x.

Problems

1. Let us evaluate by contour integration the integral $\int_{0}^{2\pi} \frac{d\theta}{1-2a\sin\theta+a^2}, 0<a<1.$

Solution:

Let,

$$z = e^{i\theta} \Rightarrow \frac{dz}{d\theta} = e^{i\theta}.i = iz$$

$$d\theta = \frac{dz}{iz}$$

$$\sin\theta = \frac{z - 1/z}{2i} = \frac{z^2-1}{2iz}.$$

$$1-2a\sin\theta+a^2 = 1-2a\left(\frac{z^2-1}{2iz}\right)+a^2$$

$$= \frac{iz - a(z^2-1) + a^2 iz}{iz}$$

$$= \frac{iz - az^2 + a + a^2 iz}{iz}$$

$$= \frac{-az^2 + z(i + a^2 i) + a}{iz}$$

$$= \frac{-az^2 + z(1+a^2)i + a}{iz}.$$

Since $z = e^{i\theta}$ C is the circle $|z| = 1$.

$$\therefore I = \int_0^{2\pi} \frac{d\theta}{1 - 2a\sin\theta + a^2} = \int_C \frac{(iz)}{-az^2 + z(1+a^2)i + a} \cdot \frac{dz}{(iz)}$$

WKT $f(z)dz = 2\pi i$ [sum of residue of f(z) at its poles]

To find poles of $\frac{1}{az^2 - i(1+a^2)z - a}$

Solving,

$$az^2 - i(1+a^2)z - a = 0.$$

$$z = \frac{(1+a^2)i \pm \sqrt{\left[-1(1+a^2)\right]^2 - 4(a)(-a)}}{2a}$$

$$= (1+a^2)i \pm \frac{\sqrt{-(1+a^4+2a^2)+4a^2}}{2a}$$

$$= \frac{(1+a^2)i \pm \sqrt{-1-a^4-2a^2+4a^2}}{2a}$$

$$= \frac{(1+a^2)i \pm \sqrt{-1-a^4-2a^2}}{2a}$$

$$\therefore I = \int_C \frac{dz}{a\left(z - \frac{i}{a}\right)(z - ia)}$$

Let,

$$f(z) = \frac{1}{a\left(z - \frac{i}{a}\right)(z - ia)}$$

$$= \frac{(1+a^2)i \pm \sqrt{-(1+a^4+2a^2)}}{2a}$$

$$= \frac{(1+a^2)i \pm \sqrt{-(1-a^2)^2}}{2a}$$

$$z = \frac{(1+a^2)i \pm (1-a^2)i}{2a}$$

$$z = \frac{(1+a^2)i + (1-a^2)i}{2a} \text{ or } \frac{(1+a^2)i - (1-a^2)i}{2a}$$

$$= \frac{i[1+a^2+1-a^2]}{2a} \text{ or } \frac{i[1+a^2-1-a^2]}{2a}$$

$$= i\left(2/_{2a}\right) \text{ or } i\left[\frac{2a^2}{2a}\right]$$

$$\frac{i}{a} \text{ or } ia$$

$$\therefore I = -\int_C \frac{dz}{a\left(z - i/_a\right)(z - ia)}$$

Let,

$$f(z) = \frac{1}{a\left(z - i/_a\right)(z - ia)}$$

∴ Poles are i/a and ia are simple poles.

$z = i/a$ lies outside C and $z = ia$ lies inside C since $|z| < 1$ and $0 < a < 1$.

$$[\text{Res of } f(z)]_{z = i/a} = 0$$

$$[\text{Res of } f(z)]_{z = ia} = \underset{z \to ia}{Lt}(z - ia) f(z),$$

$$= \underset{z \to ia}{Lt}(z - ia) \cdot \frac{1}{a(z - i/a)(z - ia)}$$

$$= \frac{1}{a(ia - i/a)} = \frac{1}{ai(a - 1/a)} = \frac{1}{ai\left(\frac{a^2 - 1}{a}\right)}$$

$$= \frac{1}{i(a^2 - 1)}$$

$$\therefore I=-\int_C \frac{dz}{a(z-i/a)(z-ia)}$$

$$=-2\pi i\left[0+\frac{1}{i(a^2-1)}\right]$$

$$=\frac{-2\pi i}{i(a^2-1)}=\frac{2\pi}{1-a^2}.$$

2. Let us evaluate $\int_{-\infty}^{\infty}\frac{x^2-x-2}{x^4+10x^2+9}dx$ using contour integration.

Solution:

Given:

$$\int_{-\infty}^{\infty}\frac{x^2-x-2}{x^4+10x^2+9}dx$$

Consider $f(z)=\frac{z^2-z+2}{z^4+10z^2+9}=\frac{z^2 \quad -z+2}{(z^2+1)(z^2+9)}$

$\int_C f(z)dz$ where C is the contour consisting of the segment of the real axis from - R to R and the semicircle Γ above the real axis having this segment as diameter.

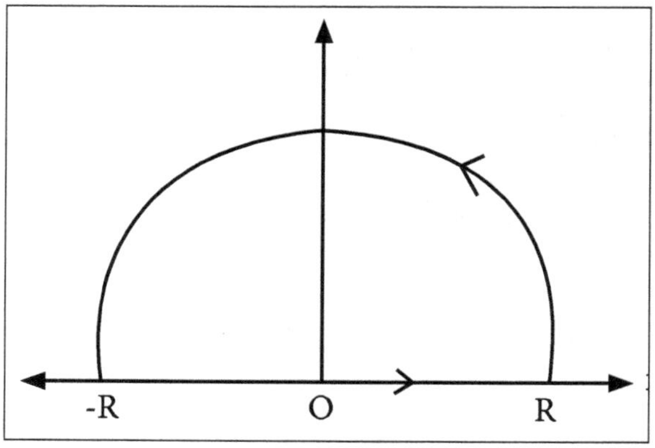

The singularities of the integer and are given by $(z^2+1)(z^2+9)=0$.

i.e., $z=\pm i, z=\pm 3i$ which are simple pole.

- The pole z = i, 3i lies inside C.
- The pole z = - i, - 3i lies outside C.

By Cauchy's Residue theorem,

$$\int_C f(z)dz = 2\pi i\left[\underset{z=1}{\text{Res } f(x)} + \underset{z=3i}{\text{Res } f(x)}\right] \qquad ...(1)$$

$$\underset{z=1}{\text{Res } f(x)} = \lim_{z\to i}(z-i)\frac{(z^2-z+2)}{(z+i)(z-i)(z^2+9)}$$

$$= \frac{1-i}{16i}$$

$$\underset{z=3i}{\text{Res } f(x)} = \lim_{z\to 3i}(z-3i)\frac{(z^2-z+2)}{(z^2+1)(z+3i)(z^2-3i)}$$

$$= \frac{(7+3i)}{48i}$$

$$\therefore \int_C f(x)dx = 2\pi i\left[\frac{(1-i)}{16i} + \frac{(7+3i)}{48i}\right]$$

$$= \frac{5\pi}{12}$$

From the diagram,

$$\int_c f(z)dz = \int_{-R}^{R} f(x)dx + \int_\Gamma f(z)dz$$

Taking limit as $\rightarrow \infty$

$$\int_\Gamma f(z)dz \rightarrow 0$$

$$= \int_{-\infty}^{\infty} f(x)dx \int_C f(z)dz$$

i.e., $\displaystyle\int_{-\infty}^{\infty} \frac{x^2-x+2}{x^4+10z^2+9}dx = \frac{5\pi}{12}$

3. Let us evaluate $\int_{-\infty}^{\infty} \frac{\cos mx}{x^2+a^2} dx$, using contour integration.

Solution:

Given:

$$\int_{-\infty}^{\infty} \frac{\cos mx}{x^2+a^2} dx,$$

Let,

$$f(z) = \frac{e^{imz}}{z^2+a^2} = \frac{e^{imz}}{(z+ai)(z-ai)}.$$

Let C be the contour, consisting of upper half of the semicircle C1: | z | = r and an interval [- r, r] on the real axis, where r is chosen sufficiently large, so that all poles lying above the x-axis are interior of C.

Now,

$$\int_C f(z)dz = \int_{C_1} f(z)dz + \int_{-r}^{r} f(z)dz$$

When $r \to \infty \int_{c_1} f(z)dz \to 0$ [by Jordan Lemma].

Hence,

$$\int_{-\infty}^{\infty} f(z)dz = \int_C f(z)dz$$

In real axis z = x

Hence,

$$\int_{-\infty}^{\infty} f(z)dz = \int_C f(z)dz$$

In real axis z = x

$$\therefore \int_{-\infty}^{\infty} f(z)dz = \int_C f(z)dz \qquad ...(1)$$

RHS of (1) can be evaluated by Cauchy's residue theorem.

By Cauchy's Residue Theorem

Hence,

$$\int_{-\infty}^{\infty} f(z)dz = \int_{C} f(z)dz$$

$$f(z) = \frac{e^{imz}}{(z+a)(z-ai)}$$

z = ai is a simple pole lies inside C.

z = – ai is a simple pole lies outside C.

$$\therefore [\text{Res of } f(z)]_{z} = -ai - 0$$

$$[\text{Res of } f(z)]_{z} = ai = \underset{z\to ai}{Lt}(z-ai)f(z)$$

$$= \underset{z\to ai}{Lt}(z-ai)\frac{e^{imz}}{(z+ai)(z-ai)}$$

$$= \underset{z\to ai}{Lt}\frac{e^{imz}}{z+ai} = \frac{e^{imz}}{2ai}$$

$$\therefore \int_{c} f(z)dz = 2\pi i\left[0 + \frac{e^{-am}}{2ai}\right] = 2\pi i\left[\frac{e^{-am}}{2ai}\right]$$

$$= \frac{\pi}{a}e^{-am}.$$

$\therefore$From (1) $\int_{-\infty}^{\infty} f(x)dx = \frac{\pi}{a}e^{-am}.$

$$\therefore \int_{-\infty}^{\infty} \frac{e^{-am}}{x^2+a^2}dx = \frac{\pi}{a}e^{-am}$$

$$\int_{-\infty}^{\infty}\left(\frac{\cos mx + i\sin mx}{x^2+a^2}\right)dx = \frac{\pi}{a}e^{-am} + 0i$$

Equating real parts,

$$\int_{-\infty}^{\infty} \frac{\cos mx}{x^2+a^2}dx = \frac{\pi}{a}e^{-am}.$$

Since,

$$\frac{\cos mx}{x^2+a^2} \text{ is even function.}$$

$$2\int_0^\infty \frac{\cos mx}{x^2+a^2}dx = \frac{\pi}{a}e^{-am}$$

$$\therefore \int_0^{2\pi} \frac{\cos mx}{x^2+a^2}dx = \frac{\pi}{2a}e^{-am}.$$

4. Let us evaluate $\int_0^{2\pi} \frac{\cos 3\theta}{5-4\cos\theta}d\theta$ using contour integration.

Solution:

Given:

$$\int_0^{2\pi} \frac{\cos 3\theta}{5-4\cos\theta}d\theta$$

Consider the unit circle C: |z| = 1, as the contour.

$$\cos\theta = \frac{z^2+1}{2z}; \cos 3\theta = \frac{1}{2}\left(z^3+\frac{1}{z^3}\right)$$

Put,

$$z = e^{i\theta}$$

$$\therefore \cos 3\theta = \frac{z^6+1}{2z^3} \;\&\; d\theta = \frac{dz}{iz}.$$

$$\therefore I = \int_0^{2\pi} \frac{\cos 3\theta}{5-4\cos\theta}d\theta = \frac{-1}{2i}\int_c \frac{z^6+1}{z^3\left(2z^2-5z+2\right)}dz$$

$$= -\frac{1}{2i}\int_c \frac{(z^6+1)}{z^3(2z-1)(z-2)}dz = \frac{1}{2i}\int_c f(z)dz.$$

We know that Cauchy's Residue theorem,

$$\int_c f(z)dz = 2\pi i \text{ [Sum of Residues of f(z) at it poles lies inside C].}$$

The poles are z = 0 of order 3 $z=\frac{1}{2}$ & $z=2$ are simple poles.

- z = 0 of order 3 lies inside C.
- z = 1/2 (simple pole) lies inside C.
- z = 2 (simple pole) lies outside C.

$$\therefore I=\left(-\frac{1}{2i}\right)(2\pi i)\left[\underset{z=0}{\text{Res}}\, f(z) \text{ of order } 3+\underset{z=1/2}{\text{Res}}\, f(z)\right] \quad ...(1)$$

Now,

$$\underset{z=0}{\text{Res}}\, f(z) \text{ of order } 3=\frac{1}{2!}\underset{z\to 0}{\text{Lt}}\frac{d^2}{dz^2}z^3\frac{(z^6+1)}{z^3(2z^2-5z+2)}$$

$$=\frac{1}{2}\underset{z\to 0}{\text{Lt}}\frac{d}{dz}\frac{(2z^2-5z+2)(6z^5)-(z^6+1)(4z-5)}{(2z^2-5z+2)^2}$$

$$=\frac{1}{2}\underset{z\to 0}{\text{Lt}}\left[\frac{(2z^2-5z+2)^2(56z^6)-(150z^5+6z^4-4)-(8z^7-25z^6+12z^5-4z+5)^2(2z^2-5z+2)(4z-5)}{(2z^2-5z+2)^4}\right]$$

$$=\frac{1}{2}\left[\frac{84}{16}\right]=\frac{21}{8}$$

$$\underset{z=1/2}{\text{Res}}\, f(z)=\underset{z\to 1/2}{\text{Lt}}\left(z-\frac{1}{2}\right)\frac{(z^6+1)}{z^3(2z-1)(z-2)} \quad \underset{z=1/2}{} f(z)=\text{Lt}_{z\to 1/2}\left(z-\frac{1}{2}\right)\frac{(z^6+1)}{z^3(2z-1)(z-2)}$$

$$=\frac{1}{2}\underset{z\to 1/2}{\text{Lt}}\frac{(z^6+1)}{z^3(z-2)}=-\frac{65}{24}.$$

Substitute in (1),

$$I=-\frac{1}{2i}(2\pi i)\left[\frac{21}{8}-\frac{65}{24}\right]$$

$$=-\pi\left[-\frac{1}{12}\right]=\frac{\pi}{12}.$$

5. Let us evaluate $\int_{-\infty}^{\infty} \frac{dx}{(x^2+1)(x^2+4)}$ using contour integration.

Solution:

Consider $f(z) = \frac{1}{(x^2+1)(z^2+4)}$; $\int_c f(z)dz$

Where C is the contour consisting of the segment of the real axis from - R to R and the semicircle F above the real axis having this segment as diameter.

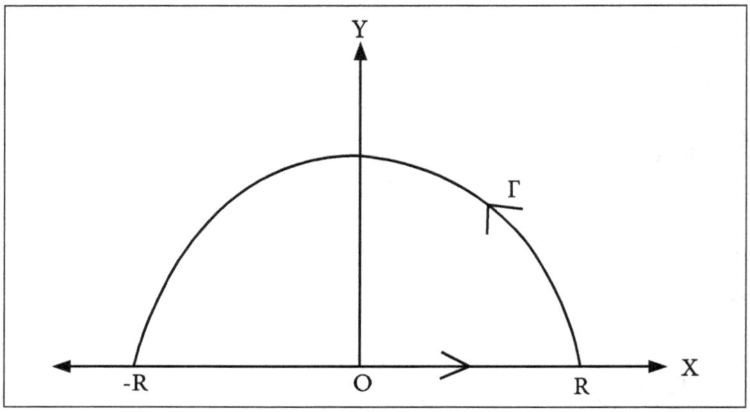

The singularities of the integral are given by,

$(z^2+1)(z^2+4) = 0 \Rightarrow z = \pm i$; $z = \pm 2i$ are simple poles

- The pole z = i; 2i which lies inside C.
- The pole z = - i; - 2i lies outside C.

∴ By Cauchy's residue theorem,

$$\int_c f(z)dz = 2\pi i\left[\underset{z=i}{\text{Res}}\, f(z) + \underset{z=2i}{\text{Res}}\, f(z)\right] \qquad ...(1)$$

$$\underset{z=i}{\text{Res}}\, f(z) = \lim_{z\to i}(z-i)\frac{1}{(z+i)(z-i)(z^2+4)} = \frac{1}{6i}$$

$$\underset{z=2i}{\text{Res}}\, f(z) = \lim_{z\to i}(z-2i)\frac{1}{(z+i)(z-i)(z^2+4)} = \frac{1}{-12i}$$

$$\therefore \int_c f(z)dz = 2\pi i\left[\frac{1}{6i} - \frac{1}{12i}\right]\frac{\pi}{6}.$$

From the diagram,

$$\int_C f(z)dz = \int_{-R}^{R} f(x)dx + \int_\Gamma f(z)dz.$$

Taking limit as $R \to \infty \int_\Gamma f(z)dz \to 0$

$$\therefore \int_{-\infty}^{\infty} f(x)dx = \int_C f(z)dz = \frac{\pi}{6}$$

$$\int_{-\infty}^{\infty} = \frac{dx}{(x^2+1)(x^2+4)} = \frac{\pi}{6}.$$

6. Let us evaluate $\int_0^{2\pi} = \frac{\cos 2\theta}{5+4\cos\theta} = d\theta$ using contour integration.

Solution:

Consider the unit circle $C: |z| = 1$.

Put,

$$z = e^{i\theta};\ \cos\theta = \frac{z^2+1}{2z} \Rightarrow \cos 2\theta = \frac{z^4+1}{2z^2}$$

$$dz = ie^{i\theta}\ d\theta$$

$$\therefore d\theta = \frac{dz}{iz}$$

$$I = \int_0^{2x} \frac{\cos 2\theta}{5+4\cos\theta} d\theta = \frac{1}{2}\int_C \frac{z^4+1}{z^2\left[5+4\left(\frac{z^2+1}{2z}\right)\right]}\frac{dz}{iz}$$

$$= \frac{1}{2i}\int_C \frac{(z^4+1)}{z^2(2z^2+5z+2)}dz$$

$$= \frac{1}{2i}\int_C \frac{(z^4+1)dz}{z^2(z+2)(2z+1)} = \frac{1}{2i}\int_C f(z)dz \qquad ...(1)$$

7. Let us know that Cauchy's Residue theorem.

Solution:

$$\int_C f(z)dz = 2\pi i \text{ [Sum of Residues of f(z) at it poles inside C].}$$

Here,

$$f(z)=\frac{z^4+1}{z^2(z+2)(2z+1)}.$$

The poles are z = 0 twice; z = - 2 and z = -1/2.

The pole z = 0; | 0 | = 0 < 1 inside c.

The pole z = - 2; |-2| = 2 >1 lies outside c.

The pole z = - 1/2; $|-\frac{1}{2}|=\frac{1}{2}<1$ lies inside c.

$$\therefore I=\frac{1}{2i}\times 2\pi i\left[\begin{matrix}\text{Res f(z)} \\ z=0 \text{ of order 2}\end{matrix}+\begin{matrix}\text{Res f(z)} \\ z=-1/2\end{matrix}\right] \qquad ...(2)$$

$$\underset{z=0}{\text{Res}}\, f(z) \text{ of order } 2 = \underset{z\to 0}{\text{Lt}}\frac{d}{dz}\frac{z^2(z^4+1)}{z^2(z+2)(2z+1)}$$

$$=\underset{z\to 0}{\text{Lt}}\frac{d}{dz}\frac{(z^4+1)}{(2z^2+5z+2)}$$

$$=\underset{z+0}{\text{Lt}}\frac{(2z^2+5z+2)(4z^3)-(z^4+1)(4z+5)}{(2z^3+5z+2)^2}$$

$$=\frac{-5}{4}.$$

$$\underset{z=-\frac{1}{2}}{\text{Res}}\, f(z) = \underset{z\to-\frac{1}{2}}{\text{Lt}}\left(z+\frac{1}{2}\right)\frac{(z^4+1)}{z^2(z+2)(2z+1)}$$

$$=\frac{1}{2}.\underset{z\to-\frac{1}{2}}{\text{Lt}}\frac{(z^4+1)}{z^2(z+2)}$$

$$= \frac{1}{2}\left(\frac{17}{6}\right) = \frac{17}{12}.$$

Substitute in (2),

$$I = \pi\left[-\frac{5}{4} + \frac{17}{12}\right] = \frac{\pi}{6}.$$

Permissions

All chapters in this book are published with permission under the Creative Commons Attribution Share Alike License or equivalent. Every chapter published in this book has been scrutinized by our experts. Their significance has been extensively debated. The topics covered herein carry significant information for a comprehensive understanding. They may even be implemented as practical applications or may be referred to as a beginning point for further studies.

We would like to thank the editorial team for lending their expertise to make the book truly unique. They have played a crucial role in the development of this book. Without their invaluable contributions this book wouldn't have been possible. They have made vital efforts to compile up to date information on the varied aspects of this subject to make this book a valuable addition to the collection of many professionals and students.

This book was conceptualized with the vision of imparting up-to-date and integrated information in this field. To ensure the same, a matchless editorial board was set up. Every individual on the board went through rigorous rounds of assessment to prove their worth. After which they invested a large part of their time researching and compiling the most relevant data for our readers.

The editorial board has been involved in producing this book since its inception. They have spent rigorous hours researching and exploring the diverse topics which have resulted in the successful publishing of this book. They have passed on their knowledge of decades through this book. To expedite this challenging task, the publisher supported the team at every step. A small team of assistant editors was also appointed to further simplify the editing procedure and attain best results for the readers.

Apart from the editorial board, the designing team has also invested a significant amount of their time in understanding the subject and creating the most relevant covers. They scrutinized every image to scout for the most suitable representation of the subject and create an appropriate cover for the book.

The publishing team has been an ardent support to the editorial, designing and production team. Their endless efforts to recruit the best for this project, has resulted in the accomplishment of this book. They are a veteran in the field of academics and their pool of knowledge is as vast as their experience in printing. Their expertise and guidance has proved useful at every step. Their uncompromising quality standards have made this book an exceptional effort. Their encouragement from time to time has been an inspiration for everyone.

The publisher and the editorial board hope that this book will prove to be a valuable piece of knowledge for students, practitioners and scholars across the globe.

Index